THE PROBLEM OF REDUCTIONISM IN SCIENCE

Episteme

A SERIES IN THE FOUNDATIONAL,
METHODOLOGICAL, PHILOSOPHICAL, PSYCHOLOGICAL, SOCIOLOGICAL, AND
POLITICAL ASPECTS OF THE SCIENCES, PURE AND APPLIED

VOLUME 18

The titles published in this series are listed at the end of this volume.

THE PROBLEM OF REDUCTIONISM IN SCIENCE

(Colloquium of the Swiss Society of Logic and Philosophy of Science, Zürich, May 18–19, 1990)

Edited by

EVANDRO AGAZZI
University of Fribourg, Switzerland
University of Genova, Italy and
President of the Swiss Society of Logic and Philosophy of Science

KLUWER ACADEMIC PUBLISHERS
DORDRECHT / BOSTON / LONDON

Library of Congress Cataloging-in-Publication Data

The problem of reductionism in science : colloquium of the Swiss Society of Logic and Philosophy of Science, Zürich, May 18-19, 1990 / edited by Evandro Agazzi.)
p. cm.
Includes bibliographical references and index.
ISBN 0-7923-1406-9 (acid free paper)
1. Science--Philosophy--Congresses. 2. Reductionism--Congresses. I. Agazzi, Evandro. II. Schweizerische Gesellschaft für Logik und Philosophie der Wissenschaften.
Q174.P82 1991
501--dc20 91-27496

ISBN 0-7923-1406-9

Published by Kluwer Academic Publishers,
P.O. Box 17, 3300 AA Dordrecht, The Netherlands.

Kluwer Academic Publishers incorporates
the publishing programmes of
D. Reidel, Martinus Nijhoff, Dr W. Junk and MTP Press.

Sold and distributed in the U.S.A. and Canada
by Kluwer Academic Publishers,
101 Philip Drive, Norwell, MA 02061, U.S.A.

In all other countries, sold and distributed
by Kluwer Academic Publishers Group,
P.O. Box 322, 3300 AH Dordrecht, The Netherlands.

Printed on acid-free paper

Printed in the Netherlands

TABLE OF CONTENTS

INTRODUCTION

The topic to which this book is devoted is *reductionism*, and not *reduction*. The difference in the adoption of these two denominations is not, contrary to what might appear at first sight, just a matter of preference between a more abstract (reductionism) or a more concrete (reduction) terminology for indicating the same subject matter. In fact, the difference is that between a philosophical doctrine (or, perhaps, simply a philosophical tenet or claim) and a scientific *procedure*. Of course, this does not mean that these two fields are separated; they are only distinct, and this already means that they are also likely to be interrelated. However it is useful to consider them separately, if at least to better understand how and why they are interconnected. Just to give a first example of difference, we can remark that a philosophical doctrine is something which makes a claim and, as such, invites controversy and should, in a way, be challenged. A scientific procedure, on the other hand, is something which concretely exists, and as such must be first of all described, interpreted, understood, defined precisely and analyzed critically; this work may well lead to uncovering limitations of this procedure, or of certain ways of conceiving or defining it, but it does not lead to really challenging it.

Let us explicitly note that this circumstance does not mean that a study of reduction does not belong to philosophy; indeed, it explicitly belongs to the philosophy of science and, more precisely, it constitutes a well established chapter of *formal* philosophy of science, in the sense that it is conceived of as one of the most relevant aspects of inter-theory relations. Within this framework, the problem of reduction may be characterized as the question of establishing when and how a certain theory T may be "reduced" to another theory T'. The answer to this question depends on several conditions, but the number (ans especially the field) of these conditions is rather restricted: it practically coincides with the way of defining what a scientific theory is. If theories are conceived of as logical formal systems (according to the original logical empiricist point of view), reduction will be conceived of in terms of purely logical relations (i.e. essentially in terms of the reciprocal logical definability of concepts, and of logical deductibility of sentences), and it will show strong affinities with explanation (understood, in the same spirit, as a particular pattern of logical deduction). If this way of conceiving theories is rejected or challenged, this may give at first the impression that the very idea of reduction is also rejected or challenged; but this is only a first (and wrong) impression. In fact, a *new way* of conceiving scientific theories replaces the old one, and this simply because a serious philosophy of science cannot eliminate the notion of scientific theory as such, but can try at most to elaborate a more adequate characterization of this notion. Once a new notion of scientific theory is proposed, it inevitably must face the problem of inter-theory relations, and this will also lead to a new way of characterizing reduction. A simple glance at the history of formal philosophy of science from the early sixties (when the logical empiricist conceptions began to undergo strong criticism) until the most recent years (when alternative approaches, like the different trends of the structuralist approach, have been developed) provides a straightforward confirmation of the above remarks.

Typical of this way of studying reduction is its "neutral" or descriptive attitude, which is in full keeping with the spirit of analytical philosophy, which has inspired, and still inspires, formal philosophy of science. This means that questions of the following kind are not considered within this approach: what reduction ought to be, why reduction should be looked at, what reduction of theories 'really means' (when it is successful and also when it is not successful), which are the ontological presuppositions or implications of reduction, and so on. All these are not formal questions, but are far from being philosophically irrelevant. Quite the contrary, they have stirred the philosophical debate from the very beginning, sometimes concealed under different terminologies, sometimes concealed under a different meaning of the word "reduction" itself. In other words, the interest about reduction has existed long before the existence of the treatment that it has received in formal philosophy of science, and this happened because the idea of reduction includes different shades of meaning, all of which have been highly appreciated by some philosophers, and strongly opposed by others throughout history.

One might maintain that the long disputes about reduction were sterile exactly because they were too deeply influenced by general philosophical tenets, and even by emotional committments, so that the pure formal treatment which is available nowadays finally allows for a philosophically rigorous treatment of the issue. Unfortunately, this is pure illusion; formal treatment simply 'isolates' a very special *aspect* of reduction, but it cannot evacuate the interest and the legitimacy of those aspects it does not take into consideration. As usual, even the opposite claim could be made: the fact that so many philosophers and scientists are still deeply interested in this issue, and discuss it without knowing anything about the sophisticated machinery of its current formal treatment, might indicate that such a treatment is simply a self-gratifying job of certain specialists, which bypasses the really important problems.

A correct position requires that both sides be taken into consideration. It is clear that behind the idea of reduction there are certain general philosophical *desiderata*, such as the search for unity, for simplification, for what is more primitive or fundamental, or for the systematic linking of knowledge, which strongly support the generalized adoption of this idea as a kind of methodological principle for our understanding of reality. The acceptance of this attitude corresponds therefore to a general philosophical choice, which can be qualified as *reductionism*. On the other hand, the existence of diversities in reality, the ability of uncovering and understanding complexity, and the fruitfulness of taking into account specificity are also philosophical *desiderata* of no less generality, which in turn would inspire the adoption of a general attitude that can be qualified as *antireductionism*.

These two tendencies have been in conflict for centuries, with the use of ontological and metaphysical arguments of various kind, but they have found new opportunities of refining their arguments when modern science became the most prestigious model of human knowledge. Indeed the growth of this science seemed to prove the fruitfulness of the reductionist point of view, and to place antireductionism in the uncomfortable position of a purely *a priori* philosophical tenet. As a reaction, it was an obvious move of the antireductionists to try to see more closely if science really supports reductionism, and this cannot be done without concretely analyzing what *reduction* really means in science, what are its features, its real achievements, and its limits. It is immediately apparent that the fulfillment of this programme gives a philosophical motivation and relevance to the strict analysis of scientific reduction which is performed in the field of formal philosophy of science, since it is within this field of investigation that many clarifications have been obtained regarding the exact features, possibilities and limitations of theory-reduction. Of course, this is not the whole story, since an historical investigation must also be undertaken for evaluating the soundness of certain specific cases of (true or alleged) reduction, while, in addition, a much more general philosophical reflection is needed in order to see which aspects of reduction are not captured by a simply formal consideration of intertheoretical relations.

To sum up, a philosophical debate on reductionism cannot remain within the framework of ontology, logic, or metaphysics, but must also take science into concrete and serious consideration. This leads to raising the question of reductionism *in science*. This question cannot be correctly tackled without relating reductionism to *reduction*, and this again implies at least two steps: first a study of the notion of reduction from a general philosophical point of view, and second, an analysis of the particular features this notion received *in science*. The present book concentrates especially on the problem of distinguishing reductionism from reduction, on an historico-philosophical determination of the concept of reduction in general, and on a philosophical analysis of the different meanings that this concept covers nowadays, expecially with reference to science. It also gives space to certain case studies of special interest, where reductionist proposals are tested in the context of different disciplines. What is not to be found here is the presentation of problems or results in the field of formal research on reduction. This is not because this research is not considered important, but simply because it needs a different kind of approach and technical preparation. In this sense, the present book is not at odds with, but rather a complement to e.g. the volume *Reduction in Science* (edited by W. Balzer, D. Pearce, H.-J Schmidt, Reidel, Dordrecht, 1984) which offers a very wide view of the formal treatment of reduction.

Let us now give a short presentation of the different papers contained in this volume. A first group of essays is of a general philosophical character. *Reductionism As Negation of Scientific Spirit* (by E. Agazzi) wants to stress the difference between reductionism and reduction. While the latter is a legitimate methodological procedure of science (if applied with enough consciousness of its limits), the former is a metaphysical absolutization of reduction, corresponding essentially to the "monistic" tenet which has been present all along the history of philosophy. Scientific knowledge, however, has characterized itself by refraining from metaphyiscal committments, and by explicitly circumcribing its investigation to specific "domains of objects". This, while legitimating a full range of reductions, does not prevent us from seeing that reduction itself is insufficient for handling complexity and accounting for the unity of complex systems. This is an additional reason for excluding any absolutization of reduction, and therefore for rejecting reductionism as contrary to the spirit of science, quite independently of the gratuitous ontological uniformity which it would try to impose on scientific inquiry.

The Power and Limits of Reduction (by M. Bunge) distinguishes and analyzes several concepts of reduction, pointing out first that reduction presupposes the concept of a level of organization and may occur either on the same level or from one level to a different one (micro-macro or macro-micro). The rapid success of microreduction has obscured the fact that in most cases it has been only partially successful, since it ignores reference to the environment of the entity under investigation (the environment belongs to a higher order level). Reduction corresponds basically to analysis, but this must be supplemented by synthesis, especially when we are studying systems and emergent properties of systems. Several examples are discussed, and the upshots are that: (a) though extremely useful, reduction is usually partial, whence "moderate reductionism" is justified; (b) the limits of reduction are rooted in the organization of the real world; whence (c) epistemology cannot ignore ontology, which obliges us to recognize the existence of systems.

The first two papers have stressed the untenability of reductionism, and shown the limitations of the most common "reductionist arguments". An interesting problem is then that of exploring whether there could be a corresponding characterization of "antireductionist arguments". It is this problem which *Theory of Antireductionist Arguments: The Bohr Case Study* (by P. Hoyningen-Huene) tackles. The paper focuses on epistemological reduction only, and reductionist, antireductionist and mediating positions are delineated. Reductionist and mediating positions express themselves through existence statements, which are argued for constructively, whereas the antireductionist position is based on impossibility claims, which are difficult to accept in science. This is why a better analysis of antireductionist arguments is needed. The strategy suggested for starting such an analysis is the study of some historically successful antireductionist arguments, with an example for the application of this strategy being proposed: the case of Bohr's argument for the irreducibility of life to physics and for considering them as "complementary". The study of this case provides a first classification of antireductionist arguments, in which Bohr's discussion finds its place (it is an argument relying upon another antireductionist argument already established in an onotlogically different domain and using analogy). In the paper it is shown that the use of the notion of complementarity for sustaining this argument is circular, since it actually presupposes the very epistemological irreducibility which it is expected to prove.

The variety of meanings of the concept of reduction, which rather often disturbs the discussions on reductionism, may receive some clarification from an historical reconstruction. This is provided in the paper *A Short History of Emergence and Reductionism* (by M. Stöckler). While "reduction" is an old notion, "reductionism" was introduced only toward the middle of the 20th century; but the great number of epistemological positions and technical investigations, originated by this proliferation of meanings, make it advisable to rather investigate the notion which is spontaneously considered as being the opposite of reductionism, i.e. "emergence". For certain authors (e.g. Nagel) emergence is not compatible with reduction, while for others (e.g. Bunge) it may be. Already J.S. Mill and other authors of the 19th century had drawn the distinction between "resultant" and "emergent" properties, while the "emergence of novelty" had become an important topic in the discussion of evolutionist theories in biology (C. Lloyd Morgan, 1923). Logical empiricism analyzed the concept of emergence as a relation between theories, disregarding ontological questions (Hempel-Oppenheim, 1948). This introduced however a strict link between the concept of emergence and the concept of explanation: through this move, emergence was brought back to the problem of theory-reduction (classically by Nagel, 1961). Unity of science was also involved (Oppenheim-Putnam, 1958). The author of the paper maintains that an ontological reductionism may be defended, also when a methodological reductionism (i.e. the reduction to the methods of a single science) fails. The concept of emergence has therefore a pragmatic pattern.

The Technical Problem of "Full Abstractness" as a Model for an Issue in Reductionism (by E. Engeler) may be seen as a proposal for circumventing the most debatable issues in the discussion of reductionism by means of a strictly formal pattern of a mathematical nature. Indeed the study uses an artificial context of reduction of one word to another to show a certain type of pitfall of reductionist methodology: compositionality may easily be violated. If the reduction pair is a "deep" world ("real"world) /observable world (theoretical model), then suppose two objects A and B, say two discrete input-output devices, are given: if they are fed with a sequence of possible inputs, each will produce a series of outputs, according to its inner workings, and those are the observale behaviors of A, and respectively B. Suppose a third device, C, to be given accepting either A or B as a component and devise a theory of the more complex systems C plus A, and C plus B. One might expect the reductionist prejudice to hold: reduce the complex systems C+A and C+B to their components A, B and C, use their full description and compose them in a description of the full system (compositionality). Well, Engeler proves and explains why this expectation prejudice fails. He then provides an explicit mathematical modeling of what is asked for in reductionism, i.e. the definition of a context for the concept of reduction, and the observation that a pair of maps is characteristic of reductions: the map from the "deep" world to the observable world, and the map from the description language of the one world to the description language of the other. By means of these definitions and observations, Engeler defines the concept of "full abstractness" and discusses some models of theoretical computer science.

A certain continuity with the above study may be seen in *A Neutral Reduction: Analytic Method and Positivism* (by J. Vuillemin). Through the discussion of several historical examples, from Descartes to Lagrange and Fourier, it is shown that physics has not known only the most familiar paradigm of reduction, consisting in "annexing" a certain branch to another one by expanding the explanatory power of the latter so as to include as a special compartment the statements of the first one. Physics actually uses also a second kind of reduction which, instead of trying to expand the explanatory power of a given theory (and suggesting in such a way a reducibility of the ontological status of the objects of the reduced theory to the ontology of the reducing theory), avoids any theoretical decision about the nature of things, and limits physical theories to a description of phenomena, in which common formal features may emerge. In this sense, such a discovery of formal isomorphisms (which concretely consists in the discovery of equations which are the same in very different fields of physics) may be called "neutral reduction", which is the most significant consequence of the adoption of the *analytical method* (i.e. the method consisting in the use of mathematical analysis) in physics. But the existence of these identical equations simply has the value of an *analogy* which, far from reflecting an alleged substantial unity of Nature that would justify reducing one class of phenomena to another more fundamental one, simply reflects the identity of the idealizations required for experimentation in the different cases. The last part of the paper discusses the affinities of this analytical reduction with certain expressions of positivistic phenomenalism (Comte, Mach, Duhem, Quine) and shows the merits and limits of this approach.

The synergy of historical considerations and theoretical discussion continues in the paper *Reductionism and Reduction in Logic and Mathematics* (by P. Weingartner). A detailed analysis of what is implicit in the general intuitive idea of reduction is proposed; for example, from unkown to known, from defined to primitive, from non-transparent to transparent, complex to simple, applied to general, non-unified to unified, intuitive theories to axiomatized or complete theories, etc., and this poses the problem of investigating which "direction" of the reduction is actually implicit in the efforts of reduction which have been historically proposed. This investigation begins with the reductionistic program already expressed by 17th century rationalism (in particular by Descartes and Leibniz), and continues with the study of the reductionist proposals implicit in the most important trends of foundational research in logic and mathematics developed in our century (logicism, intuitionism, Hilbert's program). The same method of inquiry is applied in the study of questions such as that of mathematical existence, substitution, and relevance. In all these cases it is shown that reduction basically follows one of the indicated directions, at least in one important aspect, if not in all aspects.

A last group of studies deals with the proposals of reduction advanced with regard to concrete scientific disciplines. The paper *Reductionism in Biology* (by R. Morchio) intends to be an account of a working biophysicist on how working scientists, in the field of biology, actually perceive the problem of reductionism. The concept of reductionism does not result in being an operational instrument of research for them, and is rather susceptible of deviating the attention from the real problems of their science. "Formal reductionism" (i.e. essentially inter-theoretic reductionism) is intrinsically untenable, "factual reductionism" (which is rather close to the idea of ontological reductionism) is a sort of faith with no impact on real research, and "methodological reductionism" (i.e., the prescription to adopt in biology the empirical methods used in physics and in general in the natural empirical sciences) is a useless truism for biology.

A much more severe judgment is expressed in the paper *Reductionism: Palaver Without Precedent* (by H. Primas). Reductionism is rejected on the basis of purely scientific considerations, by showing that there is no single scientifically interesting example in which at least one of the reduction schemes advocated by reductionists does really apply. In doing this, several examples taken from the best known literature are discussed and shown to be scientifically untenable. The subsequent discussion rejects reducibility of chemistry to physics, and shows that emergence of new properties is due to the braking of symmetries, which is by no means a natual consequence, or something implicit, in the given structure. The conclusion is that natural science is pluralistic, and must avail itself of several autonomous languages. Inter-theoretic relations are certainly very important, but have nothing to do with the idea of reductionism.

Two papers are devoted to the question of artificial intelligence. *Must a Science of Artificial Intelligence Be Necessarily Reductionist*? (by G.G. Granger), after the presentation of reductionism, examines the basic models of artificial intelligence and their possible reductionistic interpretation. The conclusion of this analysis is that, in spite of practical success in the simulation of several human intelligent performances, this success is still unable to produce models of the act of symbolization and, therefore, to artificially explain this fundamental capability of human intelligence. In this sense, artificial intelligence is unable to reduce natural intelligence to material, energetic and algorithmic procedures.

P. Suppes answers negatively the question asked in the title of his paper: *Can Psychological Software be Reduced to Physiological Hardware?* He follows four lines of argument. Concerning the nature of computation he argues, first, that physical location is typical of computation in a digital computer, but location of mental computations in terms of neurons seems totally out of reach. Secondly, in goal-oriented behavior physiology relates to psychology the way logic relates to physics, and he argues that no one would say that we can reduce physics to logic just because we need logic to formulate physics. The third line of argument, on computational irreducibility, refers to the modern interest in chaotic systems which has provided a sense of limitations in principle about predictability and about computational reducibility of physical systems. These more recent discoveries induce one to think that mental computations of the mind are also computationally irreducible. Finally, in a paragraph on the irrelevance of physical determinism, Suppes maintains that the classical attempts to reduce psychology to physics by general deterministic arguments amount to trying to reduce the rich facts of our mental life to an unverifiable metaphysics of determinism.

On the Problem of Reducing Value-Components in Epistemic Concepts (by E. Klevakina) is related to the topic of reduction in a rather special sense, i.e. in the sense in which "reduction" means to a certain extent "elimination". The paper considers the fact that most epistemic concepts (knowledge, belief, doubt, etc.) are often value-laden, in the sense that we attribute to them qualitative features, degrees, justifications, acceptability, which all depend apparently on preconceived values. It explores the possibility of eliminating these features by replacing them through more neutral and objective criteria. The result of the analysis is that such an elimination is only partially successful, but always interesting and illuminating.

The papers collected in this book are essentially the invited papers presented at a conference on *The Problem of Reductionism in the Sciences*, organized at the Federal Polytechnic of Zurich from May 18-19, 1990 by the Swiss Society of Logic and Philosophy of Science. The bringing about of this conference was made possible thanks to the support of the same Polytechnic and of the Swiss Academy of Natural Sciences, to which goes our gratitude. We also thank the University Council of the University of Fribourg for the financial help granted for the technical preparation of this volume, and Dr Giovanni Sommaruga and Edgardo Castro for their valuable help in the editorial work, and in the production of the camera-ready copy.

Evandro Agazzi
President of the Swiss Society of
Logic and Philosophy of Science

Evandro Agazzi

REDUCTIONISM AS NEGATION OF THE SCIENTIFIC SPIRIT

The polemic accent implicit in the title of this essay must be understood as a statement addressed against an absolutization or excess, namely, the absolutization of a methodological approach which, in other respects, is fully legitimate and correct in itself, provided one remains conscious of its conditions of application and limits. This approach is the perspective of reduction, of which «reductionism» (as in general for every «-ism») results in unjustified dogmatic generalization. But, because one may speak of reduction according to different meanings[1], it would be a difficult task to define reductionism by relying upon the various meanings of the concept of reduction. It is, therefore, easier to characterize the concept of reductionism directly, for it corresponds to an intellectual attitude, a genuine metaphysical vision, which is more fundamental and more ancient than the methodological approaches one could class under the epistemological notion of «reduction» and which, at bottom, is responsible for the very *absolutization* of which we have just spoken.

NATURE OF REDUCTIONISM

The most general characteristic identifiable in reductionism is that which may be called the *elimination of the difference*,

E. Agazzi (ed.), The Problem of Reductionism in Science, 1–29.

elimination consisting not so much in a search for that which might be common to two kinds of things, to two types of diversely expressed realities, but rather in claiming that the perceptible difference is but apparent, affirming that ,«in reality», one of the two kinds embraces also the other, as one of its possible forms of manifestation. Now, researching what is common to two different realities is one of the most normal intellectual activities. One could say that it is a basic factor in all of our understanding, that which permits us to pass from the pure and simple level of perception to that of concepts; but this does not imply the elimination of the differences. We simply recognize that, beside their diverse traits, things also possess various common traits; we deny nothing of that which experience teaches us, for the common traits themselves are, in a certain manner, attested by experience. To the contrary, when one denies the difference, when one removes from it any character of authentic reality, one becomes engaged in a *metaphysical* claim: first of all, because one attempts to affirm «what a certain reality is in itself», what is its «essence» (generally «hidden»), and, secondly, because this more profound nature is presented as going beyond the level of experience, even so far as to impose upon experience the implicit accusation of being a source of simple «appearances» (and this, moreover, is, in general, the sign of a *poor* metaphysics). If we consider many examples of reductionism, such as they are asserted regarding the relations between different sciences (for example, the thesis of the reducibility of psychology to biology, of biology to chemistry, of chemistry to physics) we readily encounter, more or less explicitly, this manner of thinking.[2]

At first sight, the reductionist tendency, as we have sketched it, seems to be characterized as a (perhaps risky) prolongation of the search for what is common to diverse realities. However, this is not the real issue; reductionism is, more precisely, a particular response which is given to the problem of searching for the unity of multiplicity, that is, the response that *monism* gives to this question. Of course, the unity of multiplicity is, again, one of the fundamental conditions for the intelligibility of the real, a condition without which one could neither think nor declare, for example, that a certain individual possesses different properties and yet is nonetheless one and not several; or that it is the same individual which has changed place, grown, or suffered from illness, who, in short, manifests himself differently in time, but always remaining the same individual. Now, the simplest solution presenting itself before this difficulty is that of performing such a *reductio ad unum* by postulating the existence of a unique kind of stuff, of some fundamental underlying substance, of which everything existing should be but a different manner of presentation, of appearance.

The first Greek philosophers actually began by proposing their world interpretation from this point of view, and monism has never failed since to manifest itself throughout history to the present day. That which has changed has simply been the identification of the said ultimate stuff, or fundamental substance, whose properties should suffice to explain all the differences that experience attests to us in that which exists. This stuff does not need to be conceived as a particular type of matter: it can as well be indicated in the Spirit, as was the case for the idealists who claimed to reduce matter to spirit, rather than spirit to matter. But that which has always remained unsatisfying in every form of monism is the real *ostension* of how that which is one can give rise to diversity, to plurality. This same difficulty unmercifully confronts as well the different forms of reductionism upheld in the scientific domain.

In order to avoid the difficulties of monism, the perspective of *pluralism* was introduced in philosophy at a relatively early stage, that is, the admission of a certain number of basic constituents of the universe, whose infinitely variable combinations, due to the influence of certain causes, would account, in principle, for the multiplicity of beings. This solution eliminates the difficulties proper to monism, but still lacks the strength to resolve the problem of the unity of multiplicity. Indeed, even if the fundamental elements are few, the problem of understanding how they combine to give rise to concrete beings with all their differences remains unsettled, for this problem cannot be solved by a simple juxtaposition of different constituents taken in variable proportions, but demands the explanation of their differentiated organization, to which also corresponds the appearance of highly differentiated properties. But there is another possible difficulty. Pluralism ultimately suffers from the same weaknesses as monism, if it fails to arrive at admitting a true *ontological diversification*, for if the fundamental elements, while individually distinct, are conceived as belonging to the same kind of reality, one abides in the presence of a masked monism. True pluralism demands that one admitS of a number of constituents of the real, sufficient to account for all the differences that reality shows, differences with regard to which one does not have the right to impose ontological limitations a priori.

This is why, therefore, the search for what is common does not necessarily lead to the negation of the difference, and the search for unity is neither satisfied via this negation. A confirmation of this fact comes from the consideration of the concept which has been able to lend itself in probably the most efficacious manner to the satisfaction of these two demands, the concept of *being*. Being appears as the most general of concepts and, at the same time, as the most powerful principle for the unification of the real. Nevertheless, the best representatives of the metaphysics of being have never ignored that being, as Aristotle already remarks, is said in several fashions, that it is an *analogical* concept, which means that there are real differences within the sphere of being, that all beings do not have the same ontological status, hence, that the notion of being itself does not permit us to reduce the beings of a certain ontological level to that of another. (The notion of being neither results in an elimination of differences, one of the fundamental reasons, moreover, for which it is so precious).

That which we have just laid down, to the extent that it exposes the reasons for the intellectual fascination of monism and highlights at the same time its profound insufficiencies, helps us to understand why and how in the domain of the sciences reductionism was able to develop. This latter is a kind of *applied monism*, that is, a monism delimited to the scientific field: apparently it advances no metaphysical pretension and is limited to bearing among the sciences the demand of generality and of unity. However, it is exactly the manner in which the satisfaction of these demands is sought that denounces its *surreptitiously* metaphysical character, as we have already established and will again have the occasion to verify. But we would like first to see how the very project of reductionism finds itself in conflict with the intellectual attitudes having permitted the constitution of modern science.

SOME CHARACTERISTICS OF SCIENTIFIC KNOWLEDGE

«Modern» science, in the usual sense of the term, is that which was constituted in the 17th century as the science of Nature, according to the adoption of an intellectual approach which was delineated and applied in an explicit and conscious way, in all of its fundamental elements, especially since Galileo. Amongst the new traits distinctively characterizing this science, with relation to the previous attitude with which one studied Nature, are the following.

i) The knowledge of natural objects does not aim at the penetration and grasping of their intrinsic «essence», but is satisfied to study a certain number of their «affections», that is, a certain number of their properties, which lend themselves to being isolated and relatively simply described along with the help of mathematical language. This means, on the one hand, that every hypothesis concerning structures or hidden properties attached to the sphere of observable phenomena must remain foreign to this knowledge of Nature, and, on the other hand, that this knowledge no longer displays the pretension of being exhaustive, of covering the entirety of what is true of a natural object with which it is concerned. In other words, the «affections» that it studies correspond to certain very delimited *points of view* that one opens upon natural things, without any pretension that the properties that one is able to grasp from these points of view be the most «essential», or that they be the only ones subsisting in the object and that one can study.

ii) Knowledge of the universal is not the *precondition* for knowledge of the particular, it is not the source from which one is able by deduction or application to expose the actual properties manifested by concrete events. Hence the search for what is common appears to be of secondary importance with respect to the exact determination of that which can be described in detail.

iii) An adequate knowledge of the parts does not necessitate knowledge of the whole. Here we are not referring, as in the preceding case, to the fact of *subsuming* the particular instance in the framework of concepts and principles of a general order, but rather to the fact of no longer considering as indispensable to the understanding of a particular event the reconstruction of its relations with the rest of the world, its insertion into a more complex context. One began to think rather the contrary, namely that the understanding of an event is found in a *deconstruction* of the event, in exposing the parts from which it «results» and their way of functioning.

DELIMITATION OF THE FIELD AS THE CONDITION FOR SCIENTIFICITY

The three characteristics we just sketched, after having determined the intellectual style of physics, have expanded little by little to the point of covering (with the necessary adaptations) the entire domain of the science of Nature, and finally the entire sphere of scientificity in general. Already at first sight, we note that the first two of the mentioned characteristics respond to an attitude opposed to that which we uncovered at the base of reductionism. In fact, they express the renunciation of this primary aspiration toward generality, of this search for the essence and ultimate explanations, which are typical of philosophical thought in its metaphysical aspect, and which have therefore authorized us to qualify the attitude of reductionism as «metaphysical». As regards the third trait, it may be considered as the conceptual root of the method of *reduction*, and the fact that it may be taken as one of the distinctive traits of scientific knowledge, while reductionism appears to be contrary to it, already tells us that reductionism and reduction must not be confused. But in order to better analyze these questions it would befit us to examine in greater detail the sense of the demands expressed in the first two characteristics, sense which we synthetically designate here as the proposal of a *conscious delimitation* of the domain of investigation.

The existence of such a delimitation as a distinctive trait of scientific research is something well known, and corresponds to the idea of *specialization*, universally recognized today as a necessary condition for this research. Nevertheless, one tends to think that specialization is a sort of necessary evil imposed upon us by quantitative reasons, that is to say, because the volume and extension of scientific knowledge in every discipline have become so enormous that no human individual is capable of mastering them all, whence the obligation that each person limit himself to cultivating and penetrating a single limited sector. Without denying what is true in this justification of specialization, we must add another which is less contingent and more structural: it directly proceeds from the Galilean program we evoked above and consists in limiting oneself, in the study of things, to the mere consideration of certain «affections», or aspects, or properties. And here lies the veritable core of specialization. The specialist is not he who knows only a very limited number of things (this would lead us to say that the ignorant man is the best specialist), but he who knows the greatest number of things *within a delimited sector*, that is to say, concerning a certain aspect of the real.

It is by thoroughly investigating this awareness, and by explicitating its conditions of application, that we arrive at a satisfactory characterization of the very idea of scientific objectivity. Having already accomplished this work on other occasions, we shall not repeat ourselves here, and we shall simply recall the essential results of this explicitation:[3]

(i) every scientific discipline (as well as every specialized sector within a discipline) is characterized by a *point of view* from which it considers reality;

(ii) the adopted point of view leads one to consider as *objects* of a discipline only those aspects of a thing accessible to that point of view;

(iii) the adopted point of view has to spell itself out in a number of specific concepts which become the *technical predicates* of the discipline in question, and whose meaning is strictly dependent upon the disciplinary context within which they fall;

(iv) among these predicates, there must be some which are connected with concrete *operations,* in virtue of which one can effectively apply the chosen point of view to the study of reality and discern whether what one affirms is true or false;

(v) thus constituted, the ensemble of predicates determines the *type of objects* with which a discipline is concerned and, in particular, the operations characterizing certain predicates determine the *ontological status* of these objects;

(vi) it follows that a single term, used in two different disciplinary contexts, receives, within the two, different meanings and refers to different types of objects;

(vii) the same *thing* may become the *object* of investigation for several different disciplines, to the extent that it lends itself to being investigated from different points of view, or (equivalently) to the extent that it possesses properties of differing nature;

(viii) none of these possible «objectivations» exhausts the thing, exactly as each objectivation corresponds to a determinate point of view, and there is an indeterminate number (even potentially infinite) of other possible points of view.

If we may consider the preceding propositions as an acceptable characterization of the modern scientific attitude or spirit, in its concern for objectivity and rigor, we can conclude that reductionism is quite opposed to such a spirit. Indeed, these propositions constitute a veritable canonization of the difference, and make of the explicitation of the differences and specificities the basic condition for the construction of scientific discourse. From this point of view, reductionism is unable to appear otherwise than the result of two possible equivocations: either the equivocation consisting in believing that one can speak about objects belonging to a certain domain in the discourse of a discipline concerned with other objects, or the equivocation consisting in believing that a determinate discipline can attain knowledge of the *thing* in its totality. The two alternatives are incompatible with the structure of scientific objectivity. The first is excluded by the fact that the predicates of a discipline determine and «construct» (by a sort of «clipping out» of things) the objects of the discipline, in such a way that these predicates are rendered constitutively incapable of speaking of «other objects» and, therefore, of «incorporating» them into their proper discipline. The most that we could expect is that the two disciplines admit of operations in common, permitting them to *refer* to the same *things*, but it always remains the case that they envisage and study different aspects in these things.

The second alternative is excluded by the fact that no investigation exhausts the «thing»: the only way to avoid this difficulty is to say that the discipline to which one wants to reduce the other or others is that which attains to the *essential* aspects of the thing or of reality, but, in this way, this discipline is made a metaphysics and no longer a science. In no way, then, can reductionism be reconciled with the scientific spirit.

REDUCTION AND THE ANALYTIC METHOD

We return now to the third characteristic of modern science, which we sketched above as being a kind of reversal of the conditions for the intelligibility of the real. While the classical tradition sought the comprehension of a thing, of a property, of an event in the possibility of situating them as elements or parts within a whole - in a totality already known and supplied with a sense - interior to which one could *reconstruct* the reasons for their existence and form of being, the new scientific perspective seeks this comprehension and this explanation, on the contrary, in the *deconstruction* of the same realities. It is these which are conceived as the totality and the whole, and one hopes to understand them through the laying out of their parts, through the observation of the manner in which these parts function, and by showing that the constitution, the properties and behavior of our (tiny) «whole» *results* from these aspects of its parts. In this consists the characterizing feature of what may be called the *analytic method.*

One could make several interesting observations with regard to this intellectual attitude: we shall limit ourselves to some very brief considerations. In the first place, one recognizes here a sort of unconscious disguise of the ancient aspiration to investigate the essence of things. This essence came, little by little, to be conceived as a hidden core, like an «intimate» substratum found in the interior of the thing, above or beneath the «affections» and its visible properties. The «intelligibility» of a thing was therefore conceived as an *intus-legere*, as a «reading inside», which was esteemed realizable thanks to a precisely «intellectual» act capable of seizing the *hidden qualities* susceptible to producing by a causal action the visible manifestations. Now this idea of «seeing inside» is found again in the idea of *deconstruction* of which we have just spoken, but in a sense which is no longer speculative or purely intellectual, but rather empirical and concrete, a bit like the child who breaks his toy in order to see what is inside. One might wonder why men had converted to this new approach, and the response (or at least a plausible response, that can be drawn from the texts of authors of that epoch) is that entire centuries had shown the sterility of the efforts to advance the knowledge of particular natural realities by means of deductions performed from the great principles supposed to assure the intelligibility of the world in its entirety, and hence applicable as well to its parts. Instead of grounding the inquiry upon the «first principles», it was hoped to better take advantage of the knowledge of certain «first realities», namely, concrete realities entering into the constitution of things falling within the sphere of our ordinary experience.

There is something more profound in this perspectival change. According to the preceding point of view, the growth of knowledge is realized by an amplification of that which is *already known* and which constitutes a sort of guarantee of the truth of that which one is going to add to that, for this latter must in some fashion «flow» from (or at least be compatible with) what one already knows. On the contrary, within the new perspective, the understanding of that which is already known is awaited from the *discovery* of that which one doesn't yet know. No one knows in advance what the analysis will reveal to us, and the intelligibility of the world depends no longer essentially upon reason but rather upon experience, subsequent experience, which, being unforeseeable, could even place into question that which we have up to present considered as known.

The simple presentation of this perspectival change suffices to make us appreciate the formidable charge it contained from the point of view of the progress of knowledge and to make us understand the new spirit which inspires it. It is the spirit of discovery, the fascination with the unknown, the search for what is new. The analytical method imposed itself not only as the fundamental method of science, but also as the intellectual style of modernity: if scientists have proceeded to a more and more elaborate *dissectio naturae* in using more and more detailed and particular «points of view», and more and more differentiated and technically refined methods of observation, the philosophers have concerned themselves with uncovering the «simple», «clear» and «distinct» ideas upon which they could build a knowledge endowed with certitude, and the mathematicians have applied themselves to the invention of instruments for the «analysis» of geometric space, of the infinitely great and of the infinitely small.

If we accept naming «reduction» the application of this analytical method (which is legitimate), we can affirm that the unprecedented success that the application of this method has assured in every domain of knowledge throughout the course of three centuries justifies to a large extent the high esteem that this approach enjoys. Nevertheless, we must more deeply examine the question in order to correctly evaluate the nature and import of these merits.

REDUCTION AND EXPLANATION

What is undeniable is the fact that the adoption of the reductive method has procured for us an enormous mass of knowledge as regards the different sets of «parts». If we want to use the same image already introduced, we can say that the adoption of this method has procured for us an unsuspected quantity of information concerning the «internal layers» of many realities which before were known to us only at their most «superficial» layer. If we take resource to another image (which is also widespread), which calls «inferior levels» that which one attains in deepening the analysis (namely, the levels to which belong the parts or components), we can say that the adoption of this method has opened for us a very rich understanding of these inferior levels. But the question now emerging is more complex: can we affirm that this augmentation of knowledge has permitted us to better *understand* the initial phenomena, namely, the phenomena of the superficial layer or of the superior level? The response depends evidently enough upon what we mean by «understanding».

The most direct sense of this notion refers to the meaning of concepts (and by this intermediary to the meaning of propositions). Now, it appears difficult to maintain that the concepts of the superior level had no meaning before the analysis or reduction, i.e., that we were unable to understand them. If indeed one had not understood them, if one truly didn't know what they meant, one would never have been able to «analyze» them and achieve any discovery of the inferior layer. Therefore, the idea of «understanding» must be conceived of in a more indirect and general sense, a sense one could express in saying that to understand something means «to attain an awareness of *how* it is». From this point of view, all new information capable of being obtained increases our understanding, for it reveals to us relations, connections, concomitances of which we previously had no precise idea: it is capable of *enriching* the meaning of the concepts at our disposal, without however changing it. This is why the use of a locution such as «superficial» layer runs the risk of producing misunderstandings: we are in a habit of attributing a negative connotation to this adjective, whereas, if we limit ourselves to an account of the simple idea of «progression» which it indicates, we can more correctly say that the «superficial» meaning is correct *within its domain of specification*. Reduction, therefore, increases our understanding in the sense just specified.

One attributes sometimes to the notion of understanding a more engaging meaning, after which to understand something means «to know *why* it is what it is». Now in this case it is more opportune to speak of explanation. In ordinary language these two concepts are used practically as synonyms, but in a more rigorous discourse it is good to distinguish them and to reserve the notion of *explanation* to the specific task of «supplying the why». Of course, even the notion of «supplying the why» is far from being conceived in a univocal manner, although in the epistemology of the sciences one practically accepts that the *ostension* of this why consists in the performance of a correct deduction. But deduction from what?

Here we encounter what is perhaps the most fundamental difference between the traditional approach and the analytic one. The traditional approach regarded the explanation of an object, of its properties and its behavior, as the possibility to deduce them from the «whole» within which it is comprised, and from the principles governing this whole and giving an order and logical connection to its parts. The analytic approach conceives explanation as a deduction beginning from the constituents and the laws which govern the relations between them. We have seen that the traditional approach had not enjoyed a great success in its efforts to explain the particular from the whole, and we must now ask ourselves if the analytic (or reductive) approach enjoys greater success. The thesis according to which this approach is effectively successful can conventionally be called *epistemological reductionism*. Is this a well founded thesis?

In order for a deduction such as it is implied by epistemological reductionism to be correctly possible, it is first of all necessary that there be a *conceptual* reduction, that is, it is necessary to be able to *translate* without residual the concepts of the superior level as functions of those of the inferior level. Next, it is necessary that all the true propositions of the superior level be transformed into theorems within the theory that describes the phenomena of the inferior level. These conditions are not easy to realize, and already from a formal logical point of view there are some well known limitations to this. However this is not in principle impossible, and an example that one could mention is probably that of the reduction of optics to electromagnetism: this «thing» which is light, and which was already «objectified» thanks to the concepts of the luminous ray and its properties in traditional optics, translates itself without residuals in the electromagnetic theory of light. Already much more complicated and controversial is the case of the possibility of reducing thermodynamics to the kinetic theory of matter.

In the case of other reductions usually proposed, the thesis of epistemological reductionism becomes more and more problematic and even false (reduction of chemistry to physics, of biology to chemistry, of psychology to biology, etc.). For example, despite all the progress of molecular biology and biochemistry, one is not in a position to remedy the absence of biological concepts in the vocabulary of chemistry, nor to give truly exhaustive definitions of the fundamental biological notions in chemical terms, which is preliminary to every enterprise which could translate all true biological propositions in as many theorems of chemistry. That which is actually being done is the assumption of the biological phenomena as a field of «things» to which the criteria of objectivation proper to chemistry are applied. This leads us to discover a whole series of *chemical aspects* of biological phenomena, but this does not yet mean reduction.

Our last affirmation permits us to say that epistemological reductionism, while not supportable as an absolute position, has the right to be applied *at the interior* of every discipline, in the following sense. Every discipline has the right to strive to read according to its criteria of objectivation any dimension of the real: the result of this effort will always be positive, for it will lead us to discover an ever more rich quantity of aspects, of constituents, of factors relevant to this discipline which are effectively present in the world. What is essential is that this unlimited effort not be transformed into the pretension of having thus *eliminated the differences*, of having erased the other points of view and fields of objectivation which derive therefrom. If one gets beyond this limit, one falls within the anti-scientific attitude we have already illustrated, and which consists exactly in the disregarding of the conditions of delimitation founding the very notion of scientific objectivity.

What is the hidden reason which promotes in general such a tendency to exceed the limits? We have already indicated it: it is *ontological reductionism*, the conception according to which there exists one single fundamental stuff of the world and that, therefore, the science which concerns itself with this fundamental stuff will arrive sooner or later to encompass the other disciplines concerned only with certain «manifestations» or more «superficial» properties of this stuff. We have already denounced the arbitrary metaphysical character of this conception, but we are now able to better see its illegitimacy from the philosophical and scientific point of view.

From the philosophical point of view, this conception introduces an arbitrary *separation* between the substance and its properties. Now, a substance is not something indeterminate and mysterious to which the properties are «added» or cling. A substance *is* the ensemble of its properties, and it is differentiated from other substances precisely *because* it possesses ontologically different properties, as likewise it may belong to the same genus as other substances if it possesses ontologically the same properties as these others, and if these properties are those which have been chosen to define the genus. Scientific objects realize this condition, so to speak, in the pure state, for they are *entirely and exclusively* defined by properties which explicitly characterize them and which are studied by the discipline concerned with them. Now, a discipline would be able to exhaust the discourse on a given substance only if this substance possessed *exclusively* the properties with which this discipline is concerned, which is impossible for at least two reasons. First of all, because a scientific discipline (as we have seen) treats only of a delimited number, generally very small, of properties of things, whereas the concrete substances (that is, the «things») effectively possess many more. In second place, because, among the properties that one attributes to things, and which can therefore be investigated by a discipline, there are several which do not belong to the substance «in itself», but insofar as it is found in relation with other substances, relations which, from the simple knowledge of «intrinsic» properties, are not predictable (and therefore unknowable). Finally, because a great number of properties can be attributed to things according to the «points of view» we *open* upon them, and these properties are again indefinite in number and unforeseeable. This is why, even admitting that a discipline be able to exhaust the properties that a substance possesses «in itself», it could never be able to master the properties arising from the two other types of relation that we have mentioned here. For example, even in the case of a tiny piece of gold, for which one might think that the description of its «intrinsic» properties is entirely furnished by physics and chemistry, one will not come to establish, physically and chemically, why it has a certain price on the market, a feature which makes it enter into the sphere of competence of the science of economics. Now, reductionist monism cultivates, in an implicit manner, the absurd persuasion that there is but one sole type of substance in the world, characterized by a small ensemble of properties whose study it attributes to a «fundamental» science, while the others would only be apparent properties added to this substance in an accidental way, and which can be «reduced» (in the sense of being ontologically eliminated) by a research which makes them «result» like appearances tied to the superficial state of our knowledge. The world, on the contrary, is full of very different substances (that is: there is not *one sole genus* of things), each one characterized by its properties, in part intrinsic, in part dependent upon the relations which they maintain with other substances, and the various disciplines attempt to study them according to limited points of view, each one occupied with only certain properties.

The preceding considerations also show us why ontological reductionism is even more untenable from a scientific point of view: taken seriously, it would mean renouncing to science in order to return to an arbitrary metaphysics, to a metaphysics which pretends to already know what is the authentic essence of things upon the basis of a partial analytical cutting-out promoted to the rank of a key for the reading of the whole. Now, even if one were willing to concede from the philosophico-metaphysical point of view that there is one sole genus of substance, this would not eliminate the legitimacy of attributing to *specific* and reciprocally irreducible investigations the study of properties that one attributes to substances as a function of their different kinds of mutual relations, there included the relations with the knowing subjects which consider them according to particular points of view.

METHODOLOGICAL REDUCTIONISM

Reductionism has been contested not only in recent times: already in the past century it was often admitted that the ontological monism which it implies is too heavy a metaphysical presupposition. Nevertheless, the necessity was put forward of a seemingly more neutral reductionism, which may be described in the following way. Let us well admit that each discipline busies itself with its specific objects. Nonetheless, we want to distinguish the disciplines that treat of their objects in a *scientific* manner from those which do so in a non-scientific manner. What does it mean to attribute to a discipline the honorable qualification «science»? The response has been: this discipline must be modeled after *the sciences* which have already merited this title, namely, after the natural sciences. This condition, in its own turn, has been explicitated in two different ways: certain authors have conceived it within the framework of epistemic reductionism, that is, in the sense that a discipline becomes scientific from the moment that it is reduced, practically, to a chapter of a natural science and, in particular, of physics, such a reduction furnishing, so to speak, a criterion for measuring the «degree of scientificity» it was capable of attaining. According to this perspective, chemistry is a bit less scientific than physics, but already well advanced upon the right path, biology - to the extent that it is inclined to be reduced to chemistry and, by this bias, to physics - appears to have good hopes to be capable of arriving some day at being strictly scientific, psychology is still quite far from this ideal, but if it should prove capable of being reduced to biology (and by this bias to chemistry and physics) it will have attained fully scientific status. And the other disciplines, such as history, law, economics, linguistics, sociology? The response has been: well, if they cannot satisfy these conditions, they shall remain outside of the frame of the sciences, though they retain interest in so far as forms of the cultural activity of men, beside the arts, morality and religion.

Another position, more tolerant and «liberal», has nevertheless manifested itself. According to this position, to be scientific, one no longer demands of a discipline that it be able to become a chapter of a natural science; it is sufficient that it adopt the *methods* of these sciences, that it be constructed after the experimental method, that it use instruments of measure, that its concepts be translated in a series of measurable magnitudes, that it have complete recourse to mathematization, that its explanations take the form of deductions performed from general laws, that it be in a position to offer verifiable predictions. Clearly, we can call *methodological reductionism* the ensemble of conditions here layed-out. But does this represent a legitimate pretension?

Against all appearances of plausibility, this pretension is not legitimate. In reality it disregards that every science is specifically occupied with its specific objects and that these *objects* are a structured ensemble of predicates dependent upon an explicitly adopted point of view. Now, it can sometimes happen that these predicates are able to satisfy the relatively complex mathematical conditions permitting us to translate them into magnitudes, but this is not obligatory, and if this is not the case, one can no longer pretend that these sciences adopt a genuine mathematization, that they express their propositions and their laws in the form of equations and disequations, that the pertinent deductions assume the form of calculations. Of course, we shall have the right to demand that these sciences offer criteria of *objectivity and rigor*, but these criteria depend on the nature of their objects. One should demand that they permit the testing of their affirmations, but this does not imply that the tests be exactly of the experimental type; they will have to furnish explanations by means of correct *arguments*, but this does not impose in advance the kind of deduction and *logic* that they must employ. In brief, while fully admitting that there is a certain «normativity» in the activity of scientific research, one must see to it that it emerges from the domain of investigation in concern, and one can not erect a method, which has been successful in a domain, as an obligatory normative criterion for all sciences: without falling into a «methodological anarchism», a «methodological pluralism» appears just as reasonable as the ontological pluralism and the epistemological pluralism we have defended up to here.[4]

THE CATEGORY OF COMPLEXITY

A certain number of the reasons we have here exposed against reductionism have already had the occasion to manifest themselves in the bosom of the science which is habitually supposed to offer us the base of all reduction, namely, physics. The crisis of the so called «classical» physics has shown that the concepts and the laws used for the edification of this physics were not everywhere applicable, which is to say that, even interior to physics, reductionism does not function. This fact has sometimes been expressed by saying that the concepts and laws valid for the world of «macro-objects» turned out not to be extendable to «micro-objects». Even if the notion of object employed in these locutions lacks the technical sense that we have given it in our context, it is all the same significant that one make allusion to it.

The reason which has led to a recognition of this change of objects is essentially the taking into consideration of the change of the *orders of magnitude*, which was at first considered in so far as a passage toward the «infinitely small», and which today has readily been recognized to be essential also when one proceeds toward the «infinitely large», that is to say, when our research objects are situated at the level of cosmic distances and time. But there are not only these two, so to speak, «categorial» differences: a third which becomes more and more clear is the difference in the *orders of complexity*: that which is valid when one treats of a simple object can easily no longer be valid when one passes to the consideration of a complex system. A rather vague perception of this problem has already existed for a long time: when we learned in school the first rudimentaries of chemistry, we were made aware of the difference between chemical «mixtures» and «combinations». In a mixture, each constituent retains all of its properties and the properties of the mixture are in a certain manner the sum or the «resultant» of the properties of the constituents (pharmaceutical research still aims today at producing such mixtures). To the contrary, in a combination one obtains a new *substance*, which has completely new properties with respect to the properties of the constituent substances (we see here that the notion of substance that we have used above is far from being a sort of archaism drawn from scholastic philosophy). Classical example, water, which possesses different properties from those of hydrogen and oxygen. Similarly, biologists have for a long time stressed that the properties, reactions and behaviors of a cell, of a tissue, of a colony of microorganisms, studied in vitro can be very different from those which they manifest in the interior of an organism.

Like the notion of order of magnitude, that of complexity is *relative* and not absolute. A grain of sand is infinitely small with respect to a mountain and infinitely large with respect to the molecules which constitute it. The solar system is infinitely large with respect to a mountain and infinitely small with respect to galactic systems. In a similar fashion, a tissue is a simple element with respect to the organism, but it is complex with respect to the cells which constitute it, and the organism is a simple element with respect to its ecosystem, just as the cell is a complex system with respect to its chemical constituents. We see then that in each of these notions enters the notion of *relation*. It is as a function of certain relations that we qualify an object as large or small, as simple or complex. Nevertheless, in the case of orders of magnitude, the relations are purely quantitative, whereas in the case of complexity they are by nature qualitative: when we say that the organism is more complex than the cell, we do not want only to indicate that the organism is «larger» (and that therefore this confers upon it certain properties which the cell has not), but that there well exist reciprocal relations between the different cells which are not the same as those which they maintain between themselves within an *in vitro* culture (case where the pure aspect of the quantitative difference would be taken into consideration); we want to underscore that the organism has certain properties *because* its cells are organized in a certain manner, which, in particular, modifies their properties and their behavior with respect to that which they do when isolated and even when they are simply «together» in a culture.

The term employed today to distinguish these types of considerations is that of *systems*, and «systems theory» is just that discipline which has developed the theoretical and practical aspects tied to this point of view, and which one could rightly credit with the merit of having tried to give a scientific status to the notion of complexity.

At first sight, one could contest this affirmation in remarking that, in reality, every traditional handbook of rational mechanics, after having devoted the necessary attention to «point mechanics», passed to the study of «system mechanics», which was developed as an application or extension of point mechanics, without need to introduce new concepts, laws or principles. The «system» was conceived as an *ensemble* of points and its properties resulted by *addition* upon these points (addition expressed by means of summations in the case of discrete systems and integrals in the case of continuous systems). Despite all of this, it does not seem arbitrary to affirm that, if one closely regards this type of theorization, one finds there such a quantity of simplifying hypotheses and idealizations, that they lead to the ignoring of the effective description of the *concretely* acting forces internal to the system and of those which act upon the system from the exterior (which are practically reduced to the consideration of their anonymous «resultants»). Concretely speaking, the actual system of which one rigorously treats in classical mechanics is that constituted of two points subject to gravitational attraction, and one knows well that from the moment that one passes to the treatment of the problem of «three bodies» this mechanics is already confronted with unavoidable difficulties (and this in being limited to the consideration of a single type of interaction, namely, gravitational interaction). Another indication in the same sense comes to us from the consideration of thermodynamics: it is well known that the second principle of this theory, the principle of increasing entropy, is applicable to *closed* systems and upon this basis it has been possible to correctly affirm that every alleged exception to this principle, which was raised in considering the thermodynamics of living beings, has no force, for these do not constitute closed systems. Nonetheless, one has the right to use this same reasoning to affirm that (precisely because of this non-closedness) thermodynamics is *not sufficient* to treat of living beings, for it does not attain the level of complexity which *objectively* characterizes them.

THE SEARCH FOR UNITY

Amongst the intellectual demands that have supported the reductionist perspective, we have already had the occasion of mentioning the search for the unity of multiplicity and we recognized that this is a legitimate demand. Have we now come to contradict this affirmation? Not necessarily: it is a question of knowing whether monism is the only possible solution to this problem. In effect, it is not so. We certainly do not want to deny that, in certain cases, the actual finding of a «common root» constitutes an objective and valid step for the partial solution to this problem, but it is not a generalizable solution. However, the considerations we have made are already sufficient to indicate to us other itineraries. The example of living organisms is already there to orient us: each organism effectively constitutes a unity, but a unity which does not deny the difference and which is even assured by the existence of the difference. If the human organism contained but brains, or kidneys, or lungs, it could not subsist: its existence as an individual entity consists in the fact that it realize an organized and finalized synthesis of different parts. If we do not fear using old notions, we can say that the unity is guaranteed by the *form*, in the profound Aristotelian sense of this concept, which means at the same time an ontological criterion of specificity, a factor of organization, and of functional and finalized coordination. If we choose to remain within the framework of Aristotelian terminology, we can say that reductionism cultivates the illusion that unity can be found in *matter*, whereas this represents precisely that which is not organized, that which is not a unity but must still receive its unification through forms. The entire swarm of substances which organic chemistry treats are constituted by a number of chemical elements that we can count on the fingers of two hands: they constitute the «matter» which, organized in different manners, gives rise to an enormous number of different organic molecules with their very diverse properties. Amongst these molecules, only a small part enters into the constitution of living matter: they are the «matter» which, again organized in very different manners, gives rise to the diverse types of cells, which in their turn are organized in the tissues and organs of organisms and so on.

Beside this way, there is another which permits the achievement of unification of multiplicity: this one attempts to find the general concepts and principles having an application in the difference, without ever suppressing it. The concepts of being, change, cause, for example, are of this type. If we use the notion of being in order to distinguish that which exists from that which does not exist, we do not claim that every being is of the same genus; if we use the notion of cause to indicate a certain type of relation between events, we are not obliged to ignore the infinity of causes which act concretely in the different types of events and so on. Moreover, a metaphysician is not obliged to imagine being as a sort of preliminary base from which arise the diverse forms of that which exists: one is not obliged to say that «first of all there is being» and then come the beings of a certain kind, and then the beings of another kind, which so appear on account of a more or less mysterious mechanism. Certainly, there are monist metaphysicians who think more or less in this way, but there are also other metaphysicians who do not follow this path. For example, even a metaphysics which admits the existence of a supreme entity that is the first cause of all that exists, affirms that this being is particular, that it is different from other existing realities, that one applies to it the notion of being in an analogical sense, just as it is analogically applied to other realities; equally, the causal act through which one attributes to this entity the creation of the world is specific and particular, different from the causality by which, for example, the artist produces his work. In brief, neither the unification which is accomplished in considering the forms unifying the multiplicity, nor the unification accomplished in using general concepts and principles, eliminates the differences and, notably, the ontological differences.

ON WHAT THERE IS

Spontaneous human understanding is implicitly convinced of the existence of this multiplicity of «ontological regions»: we think it is possible to speak with truth about trees and rocks, but also about our dreams, about numbers and mathematical functions, about our feelings, about characters from a novel, and so on. The various sciences, precisely on account of their specialization, can be considered as a very efficacious way by which man deepens his understanding in some sector of these different ontological regions. Reductionism, as we have seen, due to its monist tendency, goes against this ontological pluralism. Up to now, we have tried to understand the reason for this tendency in attributing it essentially to the desire to discover unity within the multiplicity of what exists, but the more decisive reason is probably more profound: it rests upon a certain fundamental conviction about «what really exists», a conviction which produces as a consequence the attempt to evacuate certain usually admitted forms of existence, which one wants to exclude from the domain of that which «really exists».

Every man and, in a certain sense, every culture, possesses its *Weltanschauung*, its vision of the world and of things, of which the most fundamental stratum concerns *what does exist*. There are many things of which the existence is accepted without problem (material things which strike our senses in everyday experience), but there are others for which the acceptance is very variable. For example, there were and there are cultures within which one easily admits the existence of invisible beings which populate the world around us (animism). There were ages where people admitted almost universally the existence of supernatural beings like angels and demons and attributed to them the possibility of having commerce with humans; or again men normally admitted the existence of astral influences upon human affairs and the destiny of persons. In those ages, people believed in good faith that it was possible to prove that a woman was a witch having commerce with the devil and condemn her, and it was thought that astrology is feasible as an exact science. Today, the majority of people belonging to cultures most profoundly influenced by science no longer admit such kinds of existences and, therefore, any reasoning and any evidence invoked to prove to them that a woman is a witch fails to succeed to convince them: for this reason, one would reject as impostures or hallucinations every «empirical witness» put forth as a proof of this fact, and one would look for «natural» explanations (biological, psychological or other) for those facts which could be undeniable. All this because, in the ontology of our contemporaries, there is, in general, no more space for evil spirits or at least for the type of influence of which they had before been esteemed capable.

This reduction of ontological spaces responds in part to a principle of rationality that is sometimes designated as Occam's razor (one should not multiply entities without necessity) and for another part is the consequence of certain ontological «economies», which the development of the sciences has shown to be judicious. Nevertheless, there is also a kind of counterpart of Occam's razor, which can be formulated as follows: one should not eliminate beings without necessity. Now, reductionism sins against this principle insofar as it has the tendency to eliminate ontological regions with regard to which there exists a preconceived aversion. If one *begins* with the presupposition that life does not correspond to something superior with respect to inanimate matter, if one *begins* from the presupposition that the intelligence is not something superior with respect to matter, one *intends* to prove that, and estimates that *should* be able to do it: reduction, then, appears as a way to successfully implement this program and no failure is able to make one desist from this purpose, for he will say that, *in principle*, reduction is possible, but that we still need time and research in order to complete it.

Now, as Occam's principle stands on guard against a type of dogmatism (the dogmatism of the unreasonable multiplication of ontological regions), its counterpart is also on guard against another form of dogmatism: the dogmatism of the unreasonable restriction of ontological regions. Moreover, the history of science offers us positive examples of the application of these two principles: for a long time people had believed in the existence of flogiston and of ether and then they were eliminated from the ontology of physics; during a certain period, many scientists opposed the existence of the electromagnetic field and of the atoms, but, in the end, these were admitted in the ontology of physics. The negative examples tied to the ignorance of these two principles are more persuasive if one takes them in the history of humanity: there were ages when people believed in the existence of «ontologically superior» men (kings of divine right, for example) and thanks to this amplification of ontology within the human sphere a legitimate right over the life and death of the other purely human beings was attributed to them; in more recent times, there has been the pretension to completely eliminate certain human beings from the class of men (Jews, embryos, handicapped) and from this reduction of the ontology of the human sphere has been drawn the conclusion that they may be treated as animals and even suppressed without any moral problem.

The conclusion that proceeds from the ensemble of the reflections presented in this paper could be the indication of the wisdom of an *open ontology*, which has not the pretension of knowing a priori how many ontological regions are in what is real, and which is ready to modify its framework according to the results of human research in its different sectors, without underestimating or overestimating anything. The systems-theoretic approach and the interdisciplinary attitude which today are imposing themselves more and more are already indications in this sense, and in this framework even reduction finds its place, to the extent that it exposes correlations, connections between diverse sectors and, by that, as we have observed, helps our *understanding* of complexity, without imposing the implicit metaphysicizations and the epistemological dogmatism of reductionism. After all, understanding difference while respecting its diversity helps us to better appreciate the marvellous variety and richness of what exists.

Philosophy departement-Fribourg University (Switzerland)

REFERENCES

AGAZZI, E.

(1969): *Temi e problemi di filosofia della fisica*,Milano, Manfredi, 1969, Reprint Roma, Abete,1974

(1978a): Les critères sémantiques pour la constitution de l'objet scientifique, in *La sémantique dans les sciences* (Colloque de l'Académie Internationale de Philosophie des Sciences:30 août-3septembre 1974, Rixensart), Bruxelles, Office International de Librairie, 1978,pp.13-29

(1978b): Eine Deutung der wissenschaftlichen Objektivität, in *Allgemeine Zeitschrift für Philosophie*, 3 (1978), pp.20-47

(1978c): L'objectivité scientifique est-elle possible sans la mesure?, in *Diogène*, 104 (1978),pp.96-115. Transl. in English: Is Scientific Objetivity Possible without Measurement?,in *Diogenes*, 104, pp.93-111

(1979a): Problèmes épistémologiques des sciences humaines, in *Epistemologia*, 2, special issue(1979), pp.39-66

(1979b): Analogicità del concetto di scienza. Il problema del rigore e dell'oggettività nelle scienze umane, in AA.VV. *Epistemologia e scienze umane*, Milano, Massimo, 1979, pp.57-58

(1980): L'epistemologia contemporanea: il concetto attuale di scienza, in AA.VV. *Scienza e filosofia oggi*, Milano, Massimo, 1980, pp.7-20

(1988): L'objectivité scientifique, in E. Agazzi (ed.), *L'objectivité dans les différentes sciences/Die Objektivität in den verschiedenen Wissenschaften*, Editions Universitaires,Fribourg, 1988, pp.13-25

BUNGE, M.

(1983): *Understanding the World*, Dordrecht-Boston, Reidel, 1983

BALZER, W. - PEARCE, D.A. - SCHMIDT, H.T.

(1984): *Reduction in Science* Reidel, Dordrecht-Boston, 1984

NOTES

1. See, for example, in this regard, BUNGE (1983), as well as BALZER, PEARCE, SCHMIDT, (1984)

2. Not all reductionisms are of the type that we have just exemplified here and that is sometimes called a «downward» reductionism, since reduction proceeds toward a so to speak «inferior» level in the order of complexity. There are also examples of «upward» reductionism, as results when one reduces e.g. the explanations of all human behavior to the social structure (reduction of psychology to sociology). Finally, there are also so to speak directionless reductionisms, which limit themselves to reducing to a single fundamental core a sometimes very large variety of aspects of the real. For example, certain pretensions that have been advanced to reduce to psychoanalytic terms the understanding and explanation not only of psychic phenomena and individual behavior, but also of religion, art, and literature, of peace, war, politics and economics are forms of reductionism which are, so to speak, all-directional.

This is why it appears appropriate to situate in the negation of the difference the specific feature of reductionism.

3. For more details, we refer to AGAZZI (1969, 1978a, 1978b, 1980, 1988).

4. For a more detailed exposition of the considerations elaborated here, one can see AGAZZI (1978c, 1979a, 1979b).

Mario Bunge

THE POWER AND LIMITS OF REDUCTION

1 The reduction operation

In this paper we shall be concerned with reduction as an epistemic operation and, more precisely, a kind of analysis. We take it that to reduce A to B is to identify A with B, or to include A in B, or to assert that every A is either an aggregate, a combination or an average of B's, or else a manifestation or an image of B. It is to assert that, although A and B may appear to be very different from one another, they are actually the same, or that A is a species of the genus B, or that every A results somehow from B's--or, put more vaguely, that A "boils down" to B, or that "in the last analysis" all A's are B's.

For example, the following hypotheses are instances of reduction, whether genuine or illegitimate. The heavenly bodies are ordinary bodies satisfying the laws of mechanics; heat is random molecular motion; light beams are packets of electromagnetic waves; chemical reactions are atomic or molecular combinations, dissociations or substitutions; life processes are combinations of chemical processes; humans are animals; mental processes are brain processes; and social facts result from individual actions--or conversely.

At least four of the above hypotheses exemplify a special kind of reduction, namely microreduction. A microreduction is an operation whereby things on a macrolevel are assumed to be

E. Agazzi (ed.), The Problem of Reductionism in Science, 31–49.

either aggregates or combinations of microentities, macroproperties are assumed to result either from the mere aggregation of microproperties, or from a combination of the latter, and macroprocesses are shown to result from microprocesses. In short, microreduction is the accounting of wholes by their parts. The converse operation, whereby the behavior of individuals is explained by their place or function in a whole, may be called macroreduction . Of the two kinds of reduction the former is so much more common, successful and prestigious than the latter, that the term 'reduction' is usually taken to mean "microreduction".

The following are examples of microreduction: The magnetic field of a magnet results from the alignment of the magnetic moments of the component atoms; water is composed of molecules resulting from the combination of hydrogen and oxygen atoms; and all medium and large size plants and animals are systems of cells. Note the difference between aggregation and combination, on which more will be said below.

Whereas reduction is an epistemic operation, reductionism is a research strategy, namely the methodological principle according to which (micro)reduction is in all cases necessary and sufficient to account for wholes and their properties. On the other hand macroreductionism is usually called antireductionism . The ontological partner of (micro)reductionism is atomism, while that of antireductionism is holism. In recent years antireductionism has become one of the battle-cries of the "postmodernist" reaction against science and rationality generally.

There is no gainsaying the sensational successes of reduction in all natural and social sciences ever since Descartes formulated explicitly the program of reducing everything except the mind to mechanical entities and processes--or, as he put it, to figures et movements. (No wonder Descartes is the bête noire of the "postmoderns" whether on the right or on the left.) One only needs to recall the spectacular achievements of nuclear, atomic and molecular physics, as well as of molecular biology, to understand the popularity of reductionism among scientists, and the reluctance they feel to acknowledge that reduction may be limited after all.

True, mechanism declined since the birth of field physics in mid 19th century. (See e.g. D'Abro 1939.) It is now generally understood that mechanics is only a part of physics, whence it is impossible to reduce everything to mechanics, even to quantum mechanics. Moreover, not even the whole of physics suffices for biology--which is not to say that vitalism has been resurrected. Physicalism, one of the tenets of the Vienna Circle and the Encyclopedia of Unified Science, died with the latter three decades ago--which is not to say that a richer version of materialism is dead.

However, the sharp decline of physicalism has not been the end of reductionism. Quite the contrary, reduction is still extremely successful, and it will continue to be so because all real things happen to be either systems or components of such. To reject reduction altogether is to deprive oneself of the joy of understanding many things and processes, and of the power that this knowledge confers.

Yet, as will be seen below, reduction is not omnipotent: It has limitations. In general, analysis is not enough: Eventually it must be supplemented with synthesis. This is because the world and our knowledge of it happen to be systems rather than mere aggregates of independent units. In particular, it is often necessary to combine two or more theories or even entire research fields rather than reducing one to the other. Witness the very existence of physical chemistry, biochemistry, physiological and social psychology, bioeconomics, economic sociology, and hundreds of other interdisciplines.

In the following we shall begin by attempting to identify the ontological roots of the limitations of the reduction operation. We shall subsequently examine reduction at work in physics, chemistry, biology, psychology, and social science. This examination will exhibit the limits of reduction and the ontological roots of such limits .

2 Whoes: Aggregation and combination

There are two ways a whole may come into being: by aggregation or by combination. The accretion of dust particles or of sand grains exemplify aggregation; so do garbage dumps, water pools, clouds, unruly mobs, and columns of refugees fleeing from a disaster. What characterizes all of these wholes is the lack of a specific structure composed by strong bonds: Such wholes are not cohesive. As a consequence once formed they may break up rather easily under the action of external forces.

When two or more things get together by interacting strongly in a specific way they constitute a system, i.e. a complex thing possessing a definite structure. Atomic nuclei, atoms, molecules, crystals, organelles, cells, organs, multicellular organisms, biopopulations, human families, business enterprises and other formal and nonformal organizations are systems. They may all be said to emerge through combination or self-organization rather than aggregation, even though, once formed, some of them may grow by accretion or decline by attrition. Unlike mere aggregates, systems are more or less cohesive. However, they may break down either as a result of conflicting relations among their parts or in response to external forces.

The relevance of the distinction between aggregates and systems to the problem of reduction should be obvious. In accounting for the emergence and dismantling of aggregates we focus on their composition and environment, in particular the external stimuli that favor the aggregation (or the disaggregation) process. In this case structure matters little: a heap does not cease to be a heap if its parts exchange places. Therefore, basically we explain aggregates (and their dissagregation) in terms of their composition and environment.

On the other hand structure, in particular internal structure, is essential in the case of systems. Indeed, to account for the emergence of a system we must uncover the corresponding combination or assembly process and, in particular, the bonds or links resulting in the formation of the whole. The same holds, mutatis mutandis, for any account of its breakdown. In other words, we explain the emergence, behavior and dismantling of systems in terms not only of their composition and environment, but also of their total (internal and external) structure.

In short, we may model any system s by a triple <C(s), E(s), S(s) >, the coordinates of which designate respectively the composition (collection of parts), environment, and structure (collection of links, couplings or bonds) of s . The structure is partly internal (links among the system components) and partly external (links among system components and environmental items). I call the above the CES model. As noted previously, aggregates lack a definite internal structure; however, they do have an external structure, which is composed by their inputs and outputs. Hence they too can be described by CES models.

One of the uses of CES-type modeling is that it prevents one from falling for what may be called pseudoreduction. This is the identification of a system with its composition, as exemplified by the assertions that an ice cube is "nothing but" a bunch of H2O molecules, a living being "nothing but" a bag of atoms and molecules, and a society "nothing but" a collection of individuals. This is mistaken because it involves overlooking structure, i.e. the way the components of a system combine with one another. Shorter: Nothing-but-ism is wrong because it denies the very existence of systems.

As with things so with their properties: In this case too we must distinguish between aggregation and combination. The aggregation of things is accompanied by the aggregation of their properties. When no qualitative novelty is involved, the properties of the whole are said to be resultants. For instance, joining two liquid bodies, or two heaps of sand, results in a body the mass of which equals the sum of the partial masses. On the other hand, although a system is bound to have resultant or hereditary properties, it also has emergent ones. When a system breaks down those new properties may be said to submerge. The concepts of emergence and submergence being ontological categories, they are consistent with explanation, in particular of the reductive kind. Thus, contrary to a widespread prejudice, emergentism and reductionism are mutually compatible (Bunge 1977).

Chemical reactions and biological evolution are paragons of emergence. The keys to the former are not only the atomic composition bu also the chemical bonds of various kinds and the environment. And the keys to biological evolution are genic variations, which are molecular changes, and the environmental forces causing natural selection. In both cases the peak of the process is speciation, or the emergence of systems of new kinds.

In both cases the environment is usually described in a schematic macroscopic way. In the case of chemical reactions, more often than not only the external pressure, temperature and humidity are noted along with the kind of enclosure or the lack of it. Likewise, in the case of the emergence of new biospecies only some of the global features of the environment, in particular climate and geographical accidents, are listed. In neither case is the environment analyzed into its elementary components.

This is one of the limitations of microreduction: That it treats the environment as a whole. It might be argued that this is only a practical limitation which an omniscient being would not face. This is doubtful: Certain environmental features, such as the shape of a container and the climate of a region, are macroproperties that cannot be usefully analyzed in microphysical terms because they are properties of systems acting as wholes.

As with things and their properties so with processes, i.e. changes in properties. Two processes may run parallel courses or they may interact. For example, two chemical reactions may or may not interfere with one another, If they do, they may either compete for a given chemical or, on the contrary, they may cooperate with one another producing what each "needs" for its completion. Likewise, two children growing up together influence one another's development. In both cases it is insufficient to study each process component: We must also study how these components act upon one another and how they modify or are modified by third processes. And, while some of the items (things, properties, processes) may be described in microscopic terms, others will be described in macroscopic ones.

3 Microlevels, macrolevels and their relations

For any kind of system we may distinguish at least two levels: the macrolevel and the microlevel. The macrolevel is the kind itself, i.e. the collection of all systems sharing certain peculiar properties. The corresponding microlevel is the collection of all the components of the systems in question. (In a moment it will be seen that there may be more than one microlevel.) For example, the atomic level is the set of atoms, while the molecular level is the set of molecules. (The fact that the molecular level is composed of a number of sublevels is beside the point.) In general, a level L system is composed of things of level L-1.

The distinction between levels need not be arbitrary: It often has a real counterpart in the qualitative differences between systems and their components. However, levels are collections of things, hence concepts, not concrete things. Therefore levels cannot act upon one another. In particular, the expression 'micro-macro interaction' does not denote an interaction between micro and macro levels but an interaction between individual things belonging to a microlevel and things belonging to a macrolevel.

Actually it is only in particle physics and electrodynamics that a single microlevel need be distinguished. All the other sciences study systems or even supersystems composed by systems, so that they involve the distinction between a number of microlevels. In other words, most sciences tackle nested systems, sometimes improperly called 'hierarchies'. Think, e.g., of the human brain, with its many subsystems, every one of which is composed by neurons and glial cells, which in turn are systems composed of further systems, such as the hippocampus and the primary visual cortex.

The complexity of the real systems studied in most sciences forces us to analyze the concept of composition into as many levels as needed. In other words, given the composition C(s) of a system s, and a level L of organization (or integration), we define the L-composition of s , or $C_L(s)$, as the overlap or intersection of

C(s) and L . Hence when speaking of the microlevel we should specify exactly which one we have in mind.

All of the natural, social and mixed sciences are faced with micro/macro gaps because all of them study systems of some kind or other, and all systems have components (the micro-aspect) as well as macroproperties of their own (the macro-aspect). In many cases one knows how to solve problems concerning the microlevel or the macrolevel in question, but one does not know how to relate them. In particular, one seldom knows how to account for macrofeatures in terms of microentities and their properties and changes thereof. Consequently the microspecialists (e.g., microeconomists) and macrospecialists(e.g.,macroeconomists) outnumber the experts in gap bridging.

Every problem concerning a micro-macro relation is intrinsically difficult. This difficulty is compounded by the dearth of careful philosophical analyses of the micro-macro relations. We shall proceed to sketching such an analysis. The very first task we must perform is that of distinguishing two basic types of micro-macro relations: De re, or ontological, and dedicto, or epistemological. Ontological micro-macro relations are a particular case of the part-whole relation, whereas epistemological micro-macro relations are conceptualizations of relations between microlevels and macrolevels. Let me explain.

The assembly of two or more atoms (or molecules, or cells, or animals) into a higher level entity is a case of ontological micro-to-macro relation. Likewise, the converse process of system dismantling exemplifies ontological macro-to-micro relation. The water condensation effect of a table salt molecule, the effect of the pacemaker cells on the heart, and of a leader on an organization, are further examples of the ontological micro-to-macro relation. When a limb is cut off, its cells die, when "the sun goes down" the average kinetic energy of the air molecules decreases, and when an organization is banned all of its members suffer. These are instances of the ontological macro-to-micro relation. In all of these cases links (or bonds or couplings) between micro and macro things or processes are involved.

No such bonding relations are involved in the relations among levels of organization, because--as noted above--levels are sets, hence concepts, not concrete things or processes. For example, the formula for the entropy of a thermodynamic system in a given macrostate, in terms of the number of atomic or molecular states or configurations compatible with the given macrostate (i.e. the famous formula "$S = k \ln W$"), is a micro-to-macro relation of the epistemic type. So are the formulas for the specific heat, the conductivity, and the refractive index of a body in terms of properties of its atomic constituents. Ditto Hebb's theory of learning in terms of the reinforcement of interneuronal connections: Here the macrolevel is composed by brain subsystems ("psychons") capable of performing mental functions, and the microlevel by neurons. There is no action of neurons on the brain or on the mind: there is only a conceptual relation between two levels of organization.

4 MM, Mm, mM and mm relations

Combining the micro-macro distinction with the ontologico epistemic one we get a total of eight interlevel relations:

1. MICRO-MICRO (mm)

(a) Ontological, e.g. atomic collisions and the love bond.

(b) Epistemological e.g., quantum theories of atoms, and psychological theories ofinterpersonal relations.

2. MICRO-MACRO (mM)

(a) Ontological, e.g. the interaction between an electron and an atom as a whole; a landslide caused by a small atmospheric perturbation; a social movement triggered by a charismatic leader.

(b)Epistemological, e.g. statistical mechanics, astrophysics, and a theory of animal behavior triggered by microstimuli such as a handful of photons striking a retina.

3. MACRO-MICRO(Mm)

(a) Ontological, e.g. The action of a flood or of an earthquake on an animal, and the effect of governments on individuals.

(b) Epistemological, e.g. a theory of an intrusive measurement on a microphysical entity, and a model of the course of a ship adrift in an oceanic current.

4. MACRO-MACRO (MM)

(a) Ontological e.g. Sun-Earth interaction, and rivalry between groups of animals (e.g. families).

b) Epistemological , e.g. planetary astronomy, plate tectonics theory, and international relations models.

These distinctions are relevant to the theories of definition and of explanation. They are also relevant to the old dispute between reductionists and anti-reductionist. This philosophical controversy occurs in every science. For instance, molecular biologists fight with organismic biologists, and in the social sciences individualists fight with collectivists. Whereas reductionists claim that only *mm* and *mM* relations have explanatory power, their opponents write these off and hold that only *Mm* and *MM* relations can explain.

In our view both contenders are partially right, hence partially wrong: Since all four relations exist, all of them pose problems. In particular we need to investigate how individuals of all kinds interact (*mm*), and how they assemble forming systems (*mM*) . But we also need to know how being part of a system affects the individual (*Mm*) , and how systems affect one another (*MM*). The need for such broader research projects shows that radical reductionist are just as wrong as radical antireductionists. Hence we had better adopt the systemic approach, which embraces all four relations –and more when required.

Evidently, whenever we distinguish more than two levels we are faced with several further relations. For example, interpolating a mesolevel between a micro and a macrolevel, and adding a megalevel on top of the latter, we get 4 same-level relations plus 3 inter-level relations, thus a total of 7 without skipping levels--actually 14 when the ontological-epistemological distinction is drawn. This happens, for instance, when relations between genome, cell, whole organism, and ecosystem are investigated. Another well-known four-level distinction is that between individual, firm, cartel, and whole economy.

We wind up this section by rewording the above in terms of the kinds of proposition that can be constructed when distinguishing just two levels, a microlevel (*m*) and a macrolevel (*M*). By combining the corresponding concepts we may form propositions of four different kinds, two same-level and two inter-level ones:

1. *mm*, e.g., hypotheses concerning specific nuclear forces, interneuronal connections, and face-to-face personal relations.
2. *mM*, e.g., the formulas of statistical mechanics, genotypephenotype relations, and the macrosocial outcomes of individual actions such as voting.
3. *Mm*, e.g., Lorentz's formula for the force exerted by an external magnetic field on an electron, and the hypotheses on the influence of social structure upon individual behavior.
4. *MM*, e.g., Newton's law of gravity, the rate equations of chemical kinetics, the ecological equations on interspecific competition, and data on international conflicts.

As with propositions so with explanations. I. e.in principle an explanation may contain same-level or inter-level explanatory premises or conclusions. (See Bunge 1967.) Moreover, explanations of all four kinds are needed because ours is a world of systems, and systems must be understood on their own level as well as resulting from the assembly of smaller units and constraining the behavior of their components. (See Bunge 1979.)

In other words, ontology must be realistic and it must guide epistemology if the latter is to be useful. (This view of the dependence of epistemology upon ontology is in sharp contrast with the positivist thesis that the"logic of science",and in particular the analysis of reduction, must be free from ontological assumptions: See, e.g., Carnap 1938.) One-sided ontologies, such as individualism (atomism) and collectivism (holism), outlaw each other's explanation types and thus curtail the power of science and technology. Only a systemic ontology encourages the search for explanations of all four kinds.

In the following we shall examine a few typical examples drawn from five branches of contemporary science: physics, chemistry, biology, psychology, and sociology.

5 Microreduction in action—and at bay

The quantum theory is usually regarded as the nucleus of microphysics and moreover as the key to the reduction of macrophysical to microphysical things and processes. This view is largely correct, though only with some qualifications. Indeed, the quantum theory contains a number of concepts borrowed from macrophysics, such as those of space, time, mass, electric charge, classical linear momentum, and classical energy. Thus four of these concepts (namely x, t, p and E) occur in the most basic of all state functions, namely the plane "wave" $\Psi = A \exp [i(px - Et)/h]$. Besides, the boundary conditions, which are part of the definition of every state function, constitute schematic macrophysical representations of the environment. (Example: the condition that the state function vanishes on the surface of a container.) A more dramatic and popular case is that of inseparability: A quantum system remains a system even if its components become spatially separated. Finally, in every measurement the measuring apparatus is treated as a macrophysical system described in classical or at most semiclassical terms.(See e.g. Bohr 1958.)

What holds for quantum physics holds, a fortiori, for quantum chemistry. This discipline contains not only the above-mentioned macrophysical concepts but also certain macrochemical ones. Thus one of the accomplishments of quantum chemistry is the ab initio calculation of the equilibrium constants of chemical reactions. In classical chemistry these constants are treated as empirical parameters. In quantum chemistry they are part of the theory of chemical reactions construed as inelastic scattering (collision) processes. However, this theory presupposes the rate equations of chemical kinetics, a part of classical chemistry. Indeed, consider the problem of calculating the rate or equilibrium constant k of a chemical reaction of the type "A + B → C". The phenomenological (macrochemical) equation for the rate of formation of the reaction product C is "$\frac{dn_C}{dt} = kn_An_B$" This classical equation is not deduced but postulated when the rate constant k is calculated in quantum-theoretical terms. Hence, quantum chemistry does not follow from quantum mechanics without further ado. In other words, chemistry is not fully reducible to physics. The epistemological reduction is only partial even though the ontological one is total.(For details see Bunge 1982, Lévy 1979.)

Our third case is that of biology. Undoubtedly, the most sensational advances in contemporary biology have been inspired by the thesis that organisms are nothing but bags of chemicals, whence biology is nothing but extremely complex chemistry. (see, e.g., Bernard 1865.) But this thesis, though heuristically enormously powerful, is not completely true, as we shall attempt to show by considering the cases of genetics and of the very definition of the concept of life.

At first sight, the discovery that the genetic material is composed of DNA molecules proves that genetics has been reduced to chemistry. (See, e.g., Schaffner 1969.) However, chemistry only accounts for DNA chemistry: It tells us nothing about the biological functions of DNA--e.g., that it controls morphogenesis and protein synthesis. In other words, DNA does not perform any such functions when outside a cell. Analogue: An airline pilot does not function as such at home.

The reason is of course that the very concept of a living cell is alien to chemistry. True, the cell components are physical and chemical entities, but in the cell these components are organized in a peculiarly biological way. It is also true that every single property of a cell, except for that of being alive, is shared by some physical or chemical systems. But only living cells possess jointly the dozen or so properties that characterize living systems. Consequently biology, though based on physics and chemistry, is not fully reducible to the latter. For exemple, any system of biological classification based exclusively on the degree of difference in RNA is bound to fail for missing the supramolecular features of organisms.

What about psychology: Is it reducible to biology? Assume, for the sake of the argument, that all mental processes are brain processes. Does this entail that psychology is a branch of biology and, in particular, of neuroscience? Not quite, and this for the following reasons. First, because brain processes are influenced by social stimuli, such as words and the perception of friends or foes. Now, such psychosocial processes are studied by social psychology, which employs sociological categories, such as those of society and occupation, which are not reducible to neuroscience. A second reason is that psychology employs concepts of its own, such as those of emotion, consciousness, and personality, as well as techniques of its own, such as interrogation and deceit, which go beyond biology.

We conclude then that, even though the psychoneural identity hypothesis is a clear case of ontological reduction, and tremendously fertile to boot, psychology is not reducible to neuroscience even though it has a large overlap with it. (For details see Bunge 1990.) Shorter: ontological reduction does not imply epistemological reduction. The parallel with quantum chemistry is striking.

Our final example is that of social science. Scholars otherwise as different as the idealist philosopher Wilhelm Dilthey and the behaviorist and utilitarian sociologist George Homans have attempted to reduce social science to psychology. According to psychological reductionism every social fact is the outcome of individual actions steered by the actors's beliefs values, goals, and intentions. In this perspective neither the natural nor the social environment would play any role in constraining individual actions: all persons would be free rational agents acting so as to maximize their utilities. (See e.g. Homans 1974.)

However, even the most extreme reducionists admit that an individual is likely to act differently in different situations or circumstances. But they do not bother to analyze such situations in terms of individual thoughts and actions. Instead, they describe them in molar terms--there had been a drought, a new law prescribed this or that, the country was at war, the transportation system was overloaded, and so on. In the end, then, sociological individualists cannot possibly carry out their methodological prescription.

Social scientists are bound to formulate propositions of the forms "Individual x in situation y performs action z", and "Forced by circumstance (or institution) x, individual y did z", where the concepts of situation (or circumstance) and institution are not analyzed in psychological terms. In conclusion, although social science needs psychology it is not reducible to it. Shorter: Psychological reductionism does not work in social science.

What often does work is relating two levels without attempting to reduce one to the other, as suggested by the following elementary example. It is well known that, when the interest rates rise above a certain level, the construction industry declines. In obvious symbols, $R \Rightarrow D$. This relation between two macroeconomic variables can be explained as follows. If the interests rates rise beyond a certain level, poor people cannot afford to buy or build houses, as a consequence of which the construction industry declines. In symbols,

$$R \Rightarrow \forall x(Px \Rightarrow \neg Bx) \qquad M\text{-}m$$
$$\forall x(Px \Rightarrow \neg Bx) \Rightarrow D \qquad m\text{-}M$$
$$\therefore\ R \Rightarrow D \qquad M\text{-}M$$

In conclusion, the microreduction preached by the methodological individualists cannot work because every person is a component of several social systems that constrain individual freedom in several ways. (Shorter: Structure constrains agency —which reacts back on structure.) By the same token, *mM* and *Mm* relations are bound to be key components of social science. The fact that eminent social scientists, such as Boudon (1981) and

Coleman (1990), call themselves methodological individualists while investigating such *mM* and *Mm* social relations, should not fool the philosopher.

6 Kinds and limits of reduction

Reduction may bear on concepts, propositions, explanations, or hypothetico-deductive systems. (See bunge 1983.) To reduce a concept A to a concept B is to define B in terms of B, where B refers to a thing, property, or process on a lower (or higher) level than that of the referent(s) of A. Such a definition will be called a reductive definition. (In the philosophical literature reductive definitions are usually called 'bridge hypotheses', presumably because they are often originally proposed as hypotheses. History without analysis can be misleading.) Example: "Heat = *df* Random atomic or molecular motion". Downward reductive (i.e., microreductive) definitions may also be called bottom-up definitions. On the other hand upward reductive (i.e. macroreductive) ones may be called top down definitions. But there are also same-level definitions, such as "Light is electromagnetic radiation".

The reduction of a proposition results from replacing at least one of the predicates occurring in it with the definiens of a reductive definition. For example, the psychological proposition "X was forming a linguistic expression" is reducible to the neurophysiological proposition "X's Wernicke's area was active" by virtue of the reductive definition "Formation of linguistic expressions =*df* Specific activity of the Wernicke area".

An explanation can be said to be reductive if and only if at least one of its explanans premises is a reductive definition or a reduced proposition. For example, the explanation of the existence of a concrete system in terms of the links among its parts is of the microreductive (or bottom-up) kind. On the other hand the explanation of the behavior of a component in terms of the place or function it holds in a system is of the macroreductive (or top-down) type. Work on a car assembly line (or on the origin of life) induces explanations of the first kind, while the car mechanic (and the medic) typically resorts to explanations of the second type.

The analysis of theory reduction is somewhat more complex. Call *T1* and *T2* two theories (hypothetico-deductive systems) sharing some referents, R a set of reductive definitions, and S a set of subsidiary hypotheses not contained in either *T1* or *T2*. (However, these hypotheses must be couched in the language resulting from the union of the languages of *T1* and *T2* if they are to blend with the latter.) We stipulate that

(a) T2 is fully (or strongly) reducible to *T1* = *df* *T2* follows logically from the union of *T1* and *R* ;

(b) *T2* is partially (or weakly) reducible to *T1* =*df* *T2* follows logically from the union of *T1* , *R* and *S* .

Ray optics is strongly reducible to wave optics by way of the reductive definition "Ray =*df* Normal to wave front". In turn, wave optics is strongly reducible to electromagnetism by virtue of the reductive definition of "light" as electromagnetic radiation of wavelengths comprised within a certain interval. On the other hand the kinetic theory of gases is only weakly reducible to particle mechanics because, in addition to the reductive definitions of the concepts of pressure and temperature, the former includes the subsidiary hypothesis of molecular chaos (or random initial distribution of ,positions and velocities).

Likewise, as we saw in Section S, quantum chemistry, cell biology, psychology and social science are only weakly (partially) reducible to the corresponding lower level disciplines. We also saw that even the quantum theory contains some classical concepts as well as subsidiary hypotheses, e.g. about macrophysical boundaries, so that it does not effect a complete reduction to microphysical concepts. (More on the various kinds and aspects of reduction in Bunge 1983, 1989.)

7 Conclusions

We make bold and generalize the preceding results by stating that, while partial microreduction is often successful, full microreduction seldom is. Moreover, we hazard to explain the failures of full microreduction by the hypothesis that every real thing, except for the universe as a whole, is embedded in some higher level system or other. Hence every *mM* relation is accompanied by some *Mm* relation, and both have often *mm* or *MM* concomitants. For this reason same-level definitions and explanations must be supplemented with inter-level (in particular bottom-up and top-down) definitions or explanations. Which goes to show that, to be of any use to science and technology, epistemology must match ontology. More precisely, a realistic epistemology must be paired off to a systemic ontology.

In short, although reduction should be pushed as far as possible, its limits should be realized. Consequently, moderate reductionism is a more realistic research strategy than radical reductionism, and either is more powerful than anti-reductionism.

Foundations and Philosophy of science Unit, Mc Gill University, Montreal, (Canada)

REFERENCES

Bernard, C. (1865) *Introduction à l'étude de la médicine expérimentale*. Paris: Flammarion, 1952.

Bohr, N. (1958) *Atomic Physics and Human Knowledge* . New York: John Wiley & Sons.

Boudon, R. (1981) *The Logic of Social Action*. London, Boston and Henley: Routledge & Kegan Paul.

Bunge, M. (1967) *Scientific Research*, Vol. II. Berlin-Heidelberg-New York: Springer-Verlag.

Bunge, M. (1977) Levels and reduction. *American Journal of Physiology. Regulatory, Integrative and Comparative Physiology*. 2:75-82.

Bunge, M. (1979) *A World of Systems*. Dordrecht-Boston: Reidel.

Bunge, M. (1982) Is chemistry a branch of physics? *Zeitschrift für allgemeine Wissenschayfstheorie* XIII:209-23.

Bunge, M. (1983) *Understanding the World* . Dordrecht-Boston: Reidel.

Bunge, M. (1989) Reduktion und Integration, Systeme und Niveaus, Monismus und Dualismus. In E. Pöppel, Ed., *Gehirn und Bewusstsein* , pp. 87- 104. Weinheim: VCH .

Bunge, M. (1990) What kind of discipline is psychology: Autonomous or dependent, humanistic or scientific, biological or sociological? *New Ideas in Psychology* 8:121-137. Comments and author's reply: pp.139-188. 375-379.

Carnap, R. (1938) Logical Foundations of the Unity of Science. In *International Encyclopedia of Unified Science* , Vol. I, No. 1, p. 42-62. Chicago: University of Chicago Press.

Coleman, J. S. (1990) *Foundations of Social Theory*. Cambridge MA: Belknap Press and Harvard University Press.

D'Abro, A. (1939) *The Decline of Mechanism in Modern Physics*. New York: D. Van Nostrand.

Homans, G. C. (1974) *Social Behavior: Its Elementary Forms* , rev. ed. New York: Harcourt Brace Jovanovich.

Lévy, M. (1979) Relations entre chimie et physique. *Epistemologia* II:337-370.

Schaffner, K. (1969) The Watson-Crick model and reductionism. *British Journal for the Philosophy of Science* 20:325-348.

Paul Hoyningen-Huene

THEORY OF ANTIREDUCTIONIST ARGUMENTS: THE BOHR CASE STUDY

With respect to its title, this paper has to provide the following three things: first, an explication of what a theory of antireductionist arguments is; second, a description of the Bohr case study; and third, an explanation of the colon in the title, namely, the connection between the theory of antireductionist arguments and the Bohr case study. In the initial section, I shall try to awaken the need for a theory of antireductionist arguments. In the second section, I will discuss the type of theory which could be meant, and how we could obtain such a theory. This will explain why case studies of a certain kind are needed. In the third section, I will sketch such a case study, namely, the case study of Bohr. Finally, I will present some preliminary results concerning the theory of antireductionist arguments.

1. The Need of a Theory of Antireductionist Arguments

Let us first focus on the very idea of a theory of antireductionist arguments. In order to make this plausible or to awaken the need of such a theory, I have to enter into the larger context of reductionism. It is well-known that in the history of science and philosophy, there have been many heated and sometimes extremely long disputes about reductionism. Already Aristotle had to defend his cosmology, that postulated a

E. Agazzi (ed.), The Problem of Reductionism in Science, 51–70.

fundamental difference between the celestial and the terrestrial region, against the opposite, reductionistic idea. In modern times, there have been many similar debates, such as those concerning the reduction of the mental to the physical, of thought to the perceived, of the electrical or the thermal to the mechanical, of the living to the non-living, of the chemical to the physical, of the mathematical to the logical, of the logical to the psychological, etc.

Some of these debates have been settled, for instance the question of whether electromagnetic phenomena are mechanical in nature; there is a consensus that this is not the case. But the majority of the disputes mentioned have turned out to be amazingly long-lived. Examples include the question of the reduction of the living to the non-living, and the so-called mindbody problem. There are several reasons for the longevity of controversies about reductionism. I will now sketch a few of them, restricting myself to the cognitive realm, and omitting possible sociological or psychological reasons.

For starters, the fact that the concept of reduction is extraordinarily ambiguous is a weighty reason for much confusion in debates about it. Every rational discussion about reduction has to clarify its subject matter in the very beginning by drawing some distinctions. Of course, I cannot here address all the details and the still remaining problems of these distinctions, which have been extensively, although perhaps still not exhaustively discussed in the literature (see e.g. Ayala 1974; Ayala 1989; Fodor 1974; Hoyningen-Huene 1985; Hoyningen-Huene 1989; Nagel 1961; Nickles 1973; Oppenheim/Putnam 1958; Schaffner 1967; Schaffner 1976; Sklar 1967; Wimsatt 1979).

A first fundamental distinction concerns the question of whether or not one deals with reductions concerning essentially the same domain. For instance, reductions of successive theories deal with essentially the same domain (*pace* incommensurability) for which the reduction of special relativity to classical mechanics would be an example. Different domains are involved especially in the question of the reduction within a certain hierarchy of levels, for instance among physics, chemistry, biology, psychology, and sociology. But the different domains may also be spatially separated. The question of the reduction between the celestial and the terrestrial domains provides a case in point (for which the consensus in our culture is only a few hundred years old).

In the following, I will primarily focus on reductions concerning different domains. For these situations which concern the reduction of a domain B to another, 'lower' domain A, we may distinguish four concepts of reduction.

Ontological reduction concerns the question of whether the domains A and B are ontologically identical, i.e., whether the entities of B are constituted of the same elementary substrates with the same elementary interactions as the entities of A.

Epistemological reduction concerns two questions: whether the concepts necessary for the description of B can be redefined in an extensionally equivalent way by the concepts of A, and whether the laws governing B can be derived from those of A, supplemented by the redefinitions mentioned, by suitable initial and boundary conditions, and possibly by suitable supplementary assumptions.

Explanatory reduction concerns the question of whether for every event or process in B there is a mechanism belonging to A which explains the event or process.

Methodological reduction is best construed as a residual category; it concerns all questions that are not covered by the previous categories. For instance, heuristic research strategies may or may not be reductionistic. Further distinctions can be drawn, which are without interest in our context, however.

In the following, I shall mainly deal with questions of epistemological reduction, i.e., questions concerning the interrelations among concepts and laws of two domains. Today, any more than cursory discussion of epistemological reduction is in need of a justification, since some philosophers have argued that questions of epistemological reduction are uninteresting, and even quite misleading with respect to the actual practice of science (for example Hull 1973, p. 622; Hull 1974, pp. 12, 44; Maull 1977; Schaffner 1974; Wimsatt 1979; Wimsatt 1980). Now it seems to me that an *exclusive* occupation with questions of epistemological reduction so characteristic for the heyday of neopositivism, can indeed produce a distorted image of science. For, the variety of problems of reduction in the actual sciences can indeed not be reduced to the single question of epistemological reduction. But the opposite extreme, namely the banishment of all questions concerning epistemological reduction, seems to me mistaken, too, since in epistemological reduction the question of the conceptual distance between different branches of science is treated. Even if this question were without any scientific import (which it is not), in a philosophical perspective it would still be an interesting question.

Which positions can, in principle, be articulated with respect to epistemological reduction? Clearly, there are the positions of reductionism and anti-reductionism which assert or deny the possibility of the respective reduction. But sometimes, there is the possibility of mediating positions, namely, if from a certain perspective, A and B appear as different aspects of the same thing. For instance, the question of the relation of the terrestrial and the celestial domain has, in modern times, not been answered by a reduction in one or the other direction, nor with a demonstration of the impossibility of the reduction. Rather, classical physics provides a point of view from which the opposition between the two domains dissolves.

Speaking as abstractly as before, the articulation of reductionist, anti-reductionist, and mediating positions does not seem to raise any problems of principle. But, in fact, there is a host of nasty problems concerning the precise articulation of these positions. As soon as one asks questions about the set of laws involved, about the possibility of their logical articulation, about the sort of relations of derivation admitted, about the legitimate premises in the derivation, and other related questions, the simple opposition among the principal positions disappears. As marks of orientation in the sense of ideal types, however, these positions remain useful, as I can only assert here. At any rate, the possible transitions between these principal positions have also contributed to the confusion in the debate about reduction.

If reductionist, antireductionist, and mediating positions can be distinguished, at least as ideal types, then the question arises how one can argue in support of these positions. The main question about an argument in support of the reductionist and the mediating position, respectively, can be clearly answered, at least in principle. These positions essentially assert *existence statements*: the claim that a certain reduction or a certain mediating position, respectively, indeed exists. In the empirical sciences, such existence claims are argued for *constructively*: by the construction of the entity whose existence is claimed. (In the formal sciences, there are, under certain conditions, also other strategies for the proof of existence statements but this is without relevance here.) In other words, the reduction or the mediating position must be presented, with all the claimed accomplishments. In practice, however, such constructions can be seldom carried out in every detail, and one has to remain satisfied with a certain plausibility of the possibility of the construction, instead of the construction itself. But in principle, the necessary line of argument is clear.

The situation is quite different with respect to the antireductionist position. Here we are seeking not an argument for an existence claim, but for an *impossibility claim*: the reduction of B to A is *impossible*, or: all attempts to reduce B to A are bound to *fail*. Apparently, at least in the domain of the empirical sciences, an argument for such an universal statement can never be cogent. Here, we have obviously found an additional source for the longevity of debates about reduction: It is not at all clear how to argue successfully for an antireductionist position. Correspondingly, a good deal of the literature on reduction, both on the philosophical and on the scientific side, is somewhat unsatisfactory. Arguments may seem somehow plausible but are not fully convincing; they may seem absurd but for unknown reasons; they may seem shallow but with great factual impact. In this situation, the desire may arise better to understand antireductionist arguments themselves: not to attack or defend this or that antireductionist position, but to reflect on the arguments to be used or to be avoided thereby. In other words:

the desideratum, then, is a *theory of antireductionist arguments*, with a sense of 'theory' to be clarified.

It may be surprising that, to the best of my knowledge, there is no attempt in the literature to obtain a theory of antireductionist arguments, in spite of the strong interest in the *results* of antireductionist arguments. In structure, this situation seems similar to the situation of argumentation in general before Aristotle. There were arguments which were more or less successful, but the foundations of argumentation were obscure. It is, then, an eminently philosophical task to direct reflection to argumentation itself, i.e. to invent formal logic. Of course, the problem of arguments in general is immensely more important than the problem of antireductionist arguments, but with respect to its structure it is the same problem.

2. THE THEORY OF ANTIREDUCTIONIST ARGUMENTS

So let us move to the envisaged theory of antireductionist arguments. What is this theory supposed to achieve? It should clarify antireductionist arguments, of course. But this answer has to concretized, both in a substantive and in a methodological respect. Let us begin with the substantive aspect.

Starting from the plausible assumption that there is more than one kind of antireductionist arguments, in a first step one should distinguish the different kinds. In other words, what is needed is a typology, or a *classification of the different types of antireductionist arguments*. The second step in the theory of antireductionist arguments consists then in a *detailed analysis of each of the main types of antireductionist arguments*. For instance, the following questions have to be answered for every type of argument: Which premises are used in the argument? How do these premises lead to the conclusion, i.e., on what is the argumentative force of the argument based? How strong is the argument? Where are the limits of the argument, that is, how can the argument be invalidated? This last question is interesting both in systematical and in historical respect, since some at first historically successful antireductionist arguments have lost their persuasiveness in the course of time, and the same may happen to some contemporary, apparently successful arguments. The third part of the theory of antireductionist arguments finally consists in a *collection and analysis of common pseudo-arguments.* (Note again the parallel of these three parts of the theory of antireductionist arguments and Aristotle's conception of logic).

What ways can most usefully be pursued in order to construct a theory of antireductionist arguments? Generally speaking, the following strategy seems most promising. First, historical cases of successful antireductionist arguments must be analyzed, that is, these arguments need to be historically reconstructed as accurately as possible. Second, the sought theory can be arrived at by reflection on these cases. The critical point is thus, first of all, an appropriate choice of case studies. It is worthwhile to reflect beforehand upon the conditions these cases must fulfill in order really to be case studies for the theory of antireductionist arguments. I see two necessary and one desirable condition.

According to the first, necessary condition the case studies must be examples of *good* antireductionist arguments, since the theory sought after will primarily be a positive theory. This condition is confronted with the difficulty that we do not yet know which arguments indeed are good antireductionist arguments, since it is just one of the aims of the theory of antireductionist arguments to inform us about the quality of a particular argument. It follows that the condition mentioned has to be replaced by an - in a sense - 'external' condition which can indeed be applied. The best replacement seems to be the condition that the cases to be analyzed have to be *historically successful.* The assessment of an antireductionist argument as historically successful has to based on the historical circumstances. Here, the historical impact of the argument and the scientific status of its author are of main importance. For instance, antireductionist arguments that were formulated by eminent scientists and were, at least for some time, accepted by a majority of the respective scientific community, are good candidates for case studies.

The second necessary condition is of a more technical nature. It states that there must be enough historical data available for the historical reconstruction. This requirement concerns not only the argument itself, but also the scientific and possibly even the philosophical background which contributed to make the argument a convincing argument in its own time.

The fulfillment of a third condition is not necessary although welcome - it is useful to facilitate the work on the theory of antireductionist arguments. According to this requirement, in a later period of history the persuasiveness of the respective argument should have decreased. If this is the case, it is very likely that by just studying the historical development some insight into the limits of the respective type of argument can be gained. As I said earlier, the theory of antireductionist arguments also comprises the limits of each type of argument.

Happily enough, these three conditions are not so strong that they cannot be satisfied in practice. For the following three cases, all three conditions are actually met.

First case: Aristotle's cosmology. For Aristotle, the celestial or supralunary region is not, in the ontological sense, reducible to the terrestrial or sublunary region. By implication, it is also not reducible in the epistemological sense. The elements that constitutes the celestial sphere, the ether, is fundamentally different from the four sublunary elements. Aristotle's antireductionist argument is mainly contained in his *On the Heavens*, book I, chapters 2 and 3; of course, some of his other works have to play a role in the reconstruction of the background of the argument.

Second case: Hans Driesch's neovitalism. Driesch developed his well-known neovitalism mainly in his books *Der Vitalismus als Geschichte und als Lehre* (*Vitalism as History and as Doctrine*) of 1905 and *Die Philosophie des Organischen* (*The Philosophy of the Organic*) of 1909. He was influential on biology up to the twenties. Today, neovitalism is almost completely dead, at least in 'established' biology.

Third case: Niels Bohr's antireductionism inspired by quantum mechanics, concerning the reduction of biology to physics. This is the 'Bohr case study' mentioned in the title of this paper. No word is necessary about its author and his scientific weight. Bohr's argument, which I am going to analyze in the following section, has had an 'ironic' impact on the history of biology. This argument motivated the young physicist Max Delbrück to study genetics with essentially physical methods. Delbrück eventually became one of the principal leaders of molecular biology. But this tradition turned out to be completely reductionist, at least in the understanding of its leading protagonists.

Further possible case studies concern the relationship between Cartesianism and Newtonianism, the relationship between mechanics and electrodynamics in the 19th century, the recurrence objection against the reduction of thermodynamics to statistical mechanics, and the refutation of psychologism with respect to logic by Husserl and Frege.

This brings the more programmatic part of this paper to an end. Now we will move on to some results of this program, first to the Bohr case study.

3. THE BOHR CASE STUDY

For space reasons, in this paper I will not be able to present Bohr's antireductionist argument in all its details; rather I shall concentrate on its essentials. The most elaborate form of Bohr's argument is published in his paper «Light and Life», given as a lecture in 1932 and first published in 1933 in *Nature*. Before analyzing the argument, an explication of the assertion Bohr is arguing for is in order.

In 1927, with the advent of Heisenberg's uncertainty relations and Bohr's conception of complementarity, the development of quantum mechanics came to a temporary closure. Thereafter, Bohr considered, among other things, the possible applications of the idea of complementarity to fields outside of physics. Since his youth, Bohr was familiar with the question of the relation of physics and chemistry to biology. In his youth, the alternative was simply either vitalism or mechanicism; but with the revolution of physics through quantum mechanics, new options emerged.

Bohr's problem was whether or not the new physics, i.e., classical physics *plus* quantum mechanics, was sufficient for the explanation of typical biological phenomena. From 1931 on, but not earlier, Bohr's position was that physics and chemistry cannot explain «the peculiar functions characteristic of life», as he puts it (Bohr 1933, p. 9). This position is not identical with epistemological antireductionism, since Bohr talks about *explanations* and not about *derivations*. But his position *implies* epistemological antireductionism since Bohr subscribes to an idea of explanation that is essentially identical with the Hempel-Oppenheim model of explanation, as can be shown by an analysis of the pertinent parts of his paper. Bohr's argument is, therefore, also an argument for epistemological antireductionism, and thus for us an appropriate case study.

Bohr's argument is apparently an argument by analogy:

«Indeed, the essential non-analyzability of atomic stability in mechanical terms presents a close analogy to the impossibility of a physical or chemical explanation of the peculiar functions characteristic of life» (Bohr 1933, p. 9).

This means: As classical physics turned out to be absolutely unable to explain the stability of atoms - Bohr's own postulates of 1913 were the first answer to this challenge -, so physics and chemistry are absolutely unable to explain the characteristics of life. What is the source of this analogy? In answering this question, Bohr states the following:

«I shall [...] add that the essence of the analogy being considered is the obvious exclusiveness between such typical aspects of life as the self-preservation and the self-generation of animals, on the one hand, and the subdivision necessary for any physical analysis on the other hand. Owing to this essential feature of complementarity [...]» (Bohr 1933, p. 10).

The analogy is thus generated by an «essential feature of complementarity». Though we are thereby putting our feet on notoriously slippery ground, we can give a preliminary reconstruction of Bohr's argument as follows:

Premise 1: The explanation of the stability of atoms by classical mechanics is impossible since in this situation complementarity obtains.
Premise 2: With respect to the explanation of typical aspects of life by physics, complementarity obtains, too.
Conclusion: The explanation of typical aspects of life by physics is impossible.

Central to Bohr's argument is thus the notion of complementarity that has to be discussed first. In «Light and Life», Bohr gives an explanation of this concept. But Bohr does not introduce the concept with respect to the relation between the stability of atoms and classical mechanics, as we need it in premise 1. Rather, complementarity is introduced with respect to light, immediately after he has explained that both the wave picture and the particle picture of light are indispensable:

«Indeed, the spatial continuity of our picture of light propagation and the atomicity of the light effects are complementary aspects in the sense that they account for equally important features of the light phenomena which can never be brought in direct contradiction with one another, since their closer analysis in mechanical terms demand mutually exclusive experimental arrangements» (Bohr 1933, p. 5).

This characterization of complementarity contains three aspects.

1. The domain of the complementarity relation consists in different features of one and the same phenomenon, namely in two «equally important features» (of light).

2. These different features of the phenomenon stand in an «obvious contrast» which «present us with a dilemma of a character hitherto unknown to physics», as Bohr says a littler earlier (Bohr 1933, p. 5). What does the dilemma consist in? If one assumes, as one does in classical physics, the «objective existence of phenomena independent of the means of their observation» (Bohr 1933, p. 7), then the description of light as a wave phenomenon *and* as a corpuscular phenomenon obviously implies a contradiction. Light can, then, either be a wave or a stream of corpuscles, but not both.

3. The last aspect of complementarity shows how the contradiction is avoided. A «closer analysis in mechanical terms» reveals that the contrasting features show up only under experimental arrangements that are «mutually exclusive». In other words: there is a physical mechanism that allows to avoid the imminent contradiction, namely, since it is physically impossible to realize an experimental arrangement in which one and the same beam of light shows both particle *and* wave features at the same time. To put it quite drastically: the impossibility invoked is the same as the impossibility for the same thing to be in two different places at the same time. The physical cause for the mechanism that avoids the contradiction arises «ultimately from the ignorance of the reaction of the object on the measuring instruments inherent in any measurement» (Bohr 1933, p. 7; see also Bohr 1928, pp. 54, 62ff.). Since this is the case, the notion of phenomena that are what they are, *independently* of what an observes does, loses its meaning in atomic physics, and the notion of complementarity is an indicator for this situation:

«[T]he notion of complementarity serves to symbolize the fundamental limitation, met with in atomic physics, of the objective existence of phenomena independent of the means of their observation» (Bohr 1933, p. 7).

Let this suffice for Bohr's notion of complementarity, as he introduced it to characterize the relation between the different modes of description of light. Now it has to be clarified why and how this notion of complementarity is also pertinent to the situation mentioned in the first and the second premises of Bohr's argument.

With respect to the first premise, Bohr gives the following explanation:

«[T]he unambiguous use of the concept of stationary states stands in a similar relation of complementarity to a mechanical analysis of intra-atomic motions as do light quanta to the electromagnetic theory of radiation» (Bohr 1933, p. 6).

Thus, the three aspects of complementarity described above can also be applied to the characterization of the relation between «the unambiguous use of the concept of stationary states», i.e.

roughly quantum mechanics, and «a mechanical analysis of intraatomic motions», i.e. roughly mechanics. The given explication of complementarity thus implies that both mechanics and quantum mechanics describe important features of phenomena; that there is a contrast between a mechanical and a quantum-mechanical description; and that this contrast will not lead to contradictions since where a quantum-mechanical description is adequate, the mechanical one is not, for physical reasons, and vice versa. It is mainly this last point which interests Bohr, namely, how the contradiction between two seemingly unreconcilable but indispensable partners can be avoided. Again, it is a physical mechanism that avoids the contradictions: effects of a specifically quantum-mechanical nature cannot be analyzed mechanically since they are thereby destroyed.

Is Bohr's first premise well-grounded? This seems indeed to be the case, since if mechanics and the use of the concept of 'stationary state' are complementary to each other, they are also unreconcilable (in the sense of their simultaneous application to the same situation). Only in cases where this sort of incompatibility prevails is it legitimate to ask whether the two indispensable partners can be brought into harmony by showing that they stand in the relation of complementarity.

Now we have to move on to the second premise of Bohr's argument. Bohr states here that with respect to the explanation of typical aspects of life by physics, complementarity obtains, too. By 'physics', the whole of physics is meant, i.e., classical physics and quantum mechanics together. In arguing for his second premise, Bohr sets the scene by directing the readers attention to the following dilemma. On the one hand, there are «the wonderful features which are constantly revealed in physiological investigations and which differ so markedly from what is known of inorganic matter» (Bohr 1933, p. 9). On the other hand, ontological antireductionism, i.e., a vitalism that assumes «that a peculiar vital force, unknown to physics, governs all organic life» (ibid.) has to be rejected. This rejection has the following consequence:

«If [...] we were able to push the analysis of the mechanism of living organisms as far as that of atomic phenomena, we should not expect to find any features foreign to inorganic matter» (ibid.).

But now we are facing the dilemma: if the constituents of living and non-living matter obey the same laws, then it is not intelligible why there should be a difference in the explanatory power of physics with respect to them. Bohr describes the resolution of the dilemma as follows:

«In this dilemma it must be kept in mind, however, that the conditions in biological and physical research are not directly comparable, since the necessity of keeping the object of investigation alive imposes a restriction on the former which finds no counterpart in the latter. Thus, we should doubtless kill an animal if we tried to carry the investigation of its organs so far that we could tell the part played by the single atoms in vital functions» (ibid.).

In other words: the conditions under which an organism is an object of biology, on the one hand, and an object of physics, on the other hand, physically exclude each other. This is the typical pattern of complementarity.

Let this suffice as an account of Bohr's second premise. We can now look at the pattern of Bohr's argument. Of course, formally the argument is incorrect since a universal premise is missing. Formally correct, the argument has to read as follows:

Premise 1: Whenever complementarity prevails between A and B, it is impossible to explain B by A.
Premise 2: Between physics and the typical aspects of life, complementarity prevails.
Conclusion: The explanation of typical aspects of life by physics is impossible.

Note that the conclusion is exactly the same as in the earlier presentation of the argument. If the argument is reconstructed in this way, Bohr's discussion of the wave-particle dualism is best viewed as the *introduction* of the complementarity concept, and his discussion of complementarity between mechanics and quantum-mechanics should be regarded as an *illustration* of the new first premise. With his argument, Bohr seems, then, to demonstrate both the epistemological irreducibility of biology to physics, and to provide an explanation of how these irreducible disciplines can still both be united in the overall structure of science.

4. THE THEORY OF ANTIREDUCTIONIST ARGUMENTS: PRELIMINARY RESULTS

The first task of a theory of antireductionist arguments is, as I said earlier, a classification of antireductionist arguments. We have discussed Bohr's argument which belongs to that class of arguments that make use of an analogy to an established antireductionism in some other domain. This suggests the following preliminary classification of antireductionist arguments:

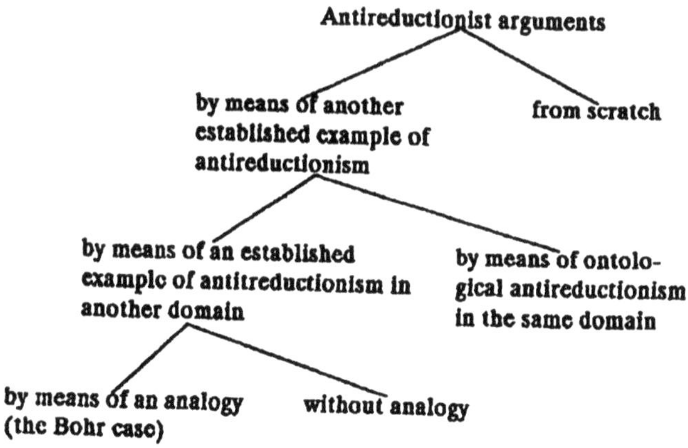

First, those antireductionist arguments that make use of another established antireductionism are distinguished from those which argue from scratch. In the first group, there are arguments that make use of an established example of antireductionism in another domain, and those that make use of an established example of antireductionism of different kind in the same domain. As far as I can see, the only candidate in the latter class is ontological antireductionism established in that domain. For instance, Frege's argument against psychologism in logic is of this type. Of course, Frege needs a separate argument to establish the ontological antireductionism in question. Finally, in the first group there are arguments that make use of analogies, and those which do not (this may turn out to be not a clear-cut distinction since the analogy arguments can perhaps all be translated into arguments that do not use analogies). At this level, we have reached the Bohr case.

The trick of this classification is, on each level, the introduction of residual classes, i.e., the introduction of the set of all possibilities so far omitted. In this way, a classification is obtained that is indeed complete although one does not really survey all cases. The danger inherent in this procedure is that further analysis of the residual classes may disclose cases which suggest a fundamentally different classification. This danger can apparently not be evaded.

What can be learned from the Bohr case study for the particular type of antireductionist argument he uses? Obviously, the diagnosis of complementarity between physics and typical biological phenomena is an essential element of this argument. Bohr seems to rest his diagnosis of complementarity mainly on its third aspect, namely, the supposed physical mechanism that avoids a possible contradiction. Now the supposition that the methods of physics are principally incompatible with the existence of specifically biological phenomena can be attacked with good reasons. Bohr realized this fact later in his life (Bohr 1963). But this is not really the essential weakness of Bohr's argument, and at the same time the weakness of all arguments that try to base irreducibility on the diagnosis of complementarity.

On what do we have to rest a correct diagnosis of complementarity in a given situation? We have to show that the three aspects of complementarity described earlier are indeed characteristics of the situation in question. This amounts to showing 1. that we have two indispensable features of some phenomenon, 2. that these features are irreconcilable with each other, and 3. that a contradiction between these features cannot occur, since for physical reason, these features cannot show up in one and the same situation. But what does it mean to demonstrate the second aspect of complementarity, i.e., that the two features are irreconcilable with each other? It means to demonstrate that these features stand in some form of contradiction, which implies that the one feature cannot be expressed in terms of the other, and vice versa. But this property is just *identical with the epistemological non-reducibility* of these features. In other words: in order to establish that in a certain situation complementarity prevails, it has to be shown that the features involved are irreducible to each other. But if this is the case, it is impossible to use complementarity in order to demonstrate that some features are epistemologically irreducible to each other, since this very fact is a necessary condition for the diagnosis of complementarity itself. In other words, any argument that tries to establish irreducibility by means of complementarity, is necessarily circular. (There are only two principle ways in which this circularity could be avoided: 1. If the third aspect of complementarity implied the second, which is not the case; see below. 2. If a criterion of complementarity could be found which is independent of its second aspect; but such a criterion is not known).

To show that for any two features of a phenomenon there is a physical mechanism that excludes the possibility that these two features show up in one and the same situation is by itself theoretically uninteresting. Even if it were true that the physicist's methods of investigation are physically incompatible with the existence of life, this would imply nothing about the theoretical relations of physics to biology. If it were impossible for some physical reason to measure the electrical conductivity of a body and its optical reflectivity in one and the same situation, this fact by itself would imply nothing about the relation between conductivity and reflectivity. This would only be interesting *if* the two other aspects of complementarity were also present, since it is mainly by this physical mechanism an imminent contradiction is avoided. Since Bohr rests his diagnosis of complementarity on the first and third aspect of complementarity alone, his argument fails.

Acknowledgement. I wish to thank Tom Rockmore for valuable suggestions about the English.

Departement of phylosophy-University of Konstanz (Germany)

REFERENCES

Asquith, P.D., Kyburg, H.E. (eds.), 1979: *Current Research in Philosophy of Science*, Ann Arbor: Edwards.

Ayala, F.J., 1974: «Introduction», in Ayala/Dobzhansky 1974: pp. VII-XVI.

Ayala, F.J., 1989: «Thermodynamics, Information, and Evolution: The Problem of Reductionism», *History and Philosophy of the Life Sciences* 11: 115-120.

Ayala, F.J., Dobzhansky, T. (eds.), 1974: *Studies in the Philosophy of Biology. Reduction and Related Problems*, Berkeley: University of California Press.

Block, N. (ed.), 1980: *Readings in Philosophy of Psychology, Vol. I.*, Cambridge: Harvard University Press.

Bohr, N., 1928: «The Quantum Postulate and the Recent Development of Atomic Theory», in Bohr 1934, pp. 52-91 (originally in *Nature (Suppl.)* 121: 580-590 (1928)).

Bohr, N., 1933: «Light and Life», in Bohr 1958: pp. 3-12 (originally in *Nature* 131: 421-423, 457-459 (1933)).

Bohr, N., 1934: *Atomic Theory and the Description of Nature*, Cambridge: Cambridge University Press. Reprint: *The Philosophical Writings of Niels Bohr, Vol. I*, Woodbridge: Ox Bow Press, 1987.

Bohr, N., 1958: *Atomic Physics and Human Knowledge*, New York: Wiley. Reprint: *The Philosophical Writings of Niels Bohr, Vol. II, Essays 1932-1957 on Atomic Physics and Human Knowledge*, Woodbridge: Ox Bow Press, 1987.

Bohr, N., 1963: «Licht und Leben - noch einmal», *Naturwissenschaften* 50: 725-727. Reprinted in Meyenn/Stolzenburg/Sexl 1985: pp. 195-202.

Fodor, J.A., 1974: «Special Sciences, or The Disunity of Science as a Working Hypothesis», in Block 1980: 120-133 (originally in *Synthese* 28: 97-115 (1974)).

Hoyningen-Huene, P., 1985: «Zu Problemen des Reduktionismus der Biologie», *Philosophia Naturalis* 22: 271-286.

Hoyningen-Huene, P., 1989: «Epistemological Reductionism in Biology: Intuitions, Explications, and Objections», in Hoyningen-Huene/Wuketits 1989: pp. 29-44.

Hoyningen-Huene, P., Wuketits, F.M. (eds.), 1989: *Reductionism and Systems Theory in the Life Sciences: Some Problems and Perspectives*, Dordrecht: Kluwer.

Hull, D.L., 1973: «Reduction in Genetics - Doing the Impossible», in Suppes 1973: pp. 619-635.

Hull, D.L., 1974: *The Philosophy of Biological Science*, Englewood Cliffs: Prentice Hall.

Maull, N., 1977: «Unifying Science Without Reduction», in Sober 1984: pp. 509-527 (originally in *Studies in History and Philosophy of Science* 9: 143-162 (1977)).

Meyenn, K.v., Stolzenburg, K., Sexl, R. (eds.), 1985: *Niels Bohr 1885-1962. Der Kopenhagener Geist in der Physik*, Braunschweig: Vieweg.

Nagel, E., 1961: *The Structure of Science. Problems in the Logic of Scientific Explanation*, London: Routledge & Kegan Paul.

Nickles, T., 1973: «Two Concepts of Intertheoretic Reduction», *Journal of Philosophy* 70: 181-201.

Oppenheim, P., Putnam, H., 1958: «Unity of Science as a Working Hypothesis», *Minnesota Studies in the Philosophy of Science* 2: 3-36.

Schaffner, K.F, 1967: «Approaches to Reduction», *Philosophy of Science* 34: 137-147.

Schaffner, K.F., 1974: «The Peripherality of Reductionism in the Development of Molecular Biology», *Journal for the History of Biology* 7: 111-139.

Schaffner, K.F., 1976: «Reduction in Biology: Prospects and Problems», in Sober 1984: pp. 428-445.

Sklar, L., 1967: «Types of Inter-Theoretic Reduction», *British Journal for the Philosophy of Science* 18: 109-124.

Sober, E. (ed.), 1984: *Conceptual Issues in Evolutionary Biology. An Anthology*, Cambridge, MA: MIT Press.

Suppes, P. (ed.), 1973: *Proceedings of the Fourth International Congress of Logic, Methodology, and Philosophy of Science* Dordrecht: North Holland.

Wimsatt, W.C., 1979: «Reduction and Reductionism», in Asquith/Kyburg 1979: pp. 352-377.

Wimsatt, W.C., 1980: «Reductionistic Research Strategies and Their Biases in the Units of Selection Controversy», in Sober 1984: pp. 142-183.

Manfred Stöckler

A SHORT HISTORY OF EMERGENCE AND REDUCTIONISM

1. Introduction

Many people mean many things when they use the words "emergence" and "reductionism". In order to understand what is meant in any particular case and in order to find out how to use these concepts in a fruitful way, we have to understand the context of problems connected with "emergence" and "reductionism". In this situation, the history of science and philosophy might help to prepare for the systematic discussions on reductionism in the sciences which follow in this conference.

My historical account will present the scientific context and some paradigmatic cases of reduction and emergent properties. Current systematic accounts of reductionism should be examined by comparing them with the examples from history. The many facets of "emergence" and "reductionism" could even help to develope a new, more uniform explication of «emergence».
I will confine to authors explicitely using the concepts "emergence" and "reductionism". Let me start with "reductionism" and postpone the more comprehensive story of "emergence".

2. The Concept of Reductionism

Problems of reduction have been discussed a long time, but the word "reductionism" was not employed before the mid-20th

E. Agazzi (ed.), The Problem of Reductionism in Science, 71–90.

century. Unlike "reductionism", "to reduce" (and "reduction") has been used for centuries. The dictionary lists various meanings;[2] for example: To bring back, to restore, to bring into another language, to change something to a different form, to bring down to a smaller amount. In chemistry it means (already in the 18th century) "to resolve a compound into simpler constituents" (the opposite of oxidation); in logic "to bring a syllogism into an equivalent form" (especially to one of the moods of the first figure).

"Reductionism" is a word of the second part of our century. The first famous author[3] writing on "reductionism" was W. V. O. Quine, who gave this name to one of his two dogmas of empiricism: «The other dogma is *reductionism*: the belief that each meaningful statement is equivalent to some logical construct upon terms which refer to immediate experience.»[4] Maybe Quine thought of Carnap's reduction sentences and of similar terms of that time like "behaviourism" or "physicalism". In both cases we have the general idea of reducing one system of propositions to another one. In a similar way R. M. Hare talked of "reductionism", when he discussed the problem of whether the language of values could be replaced by imperatives: «Nor am I committing the sin of "reductionism" ... I am not, that is to say, trying to analyse one kind of language in terms of another.»[5]

Later in the fifties "reductionism" spread into many fields. The concept was extended and no longer confined to an epistemic use. Old problems of the philosophy of biology (mechanism versus vitalism), of the philosophy of psychology (mind-body problem) and of the foundations of social sciences (methodological individualism versus collectivism) were discussed under the new heading of "reductionism". M. Scriven gave the following characterization: «The reductionists are those who have supported programs which involve "reducing" optics and chemistry to the physics of wave, fields and particles, or psychology to the study of observable behavior, or biology to chemistry, or history and ethics to psychology, or ethics to decision theory etc.»[6]

In the same 1958 volume of the Minnesota Studies in the Philosophy of Science we find the famous paper of Paul Oppenheim and Hilary Putnam "Unity of Science as a Working Hypothesis"[7]. Their reductionistic program contains four elements reflecting various aspects of reductionism and emergence:

1. A well-developed concept of reduction (as defined by Kemeny and Oppenheim), which is compatible with a part-whole relation (micro-reduction).

2. An ordering of the branches of science describing distinct levels of reality. The things on a particular level are composed of simpler elements belonging to the next lower level.

3. A program of unity of science based on the foundation of physics, the most basic level of science. This program is anti-vitalistic and all-in-all more metaphysical than epistemological.

4. The empirically-supported belief in cosmic evolution. During this evolution objects of a given level combine to form wholes belonging to a higher level. Things that appear later in time can be accounted for in terms of things and processes at earlier times.

The present discussion is characterized by two opposite directions: On one side we have a sometimes emotionally heated debate on methodological reductionism (for example in sociobiology[8]), on the other side we have sophisticated investigations of the formal properties of theory reduction.

So it seems necessary to disentangle the various aspects of reductionism by going back to the roots of the modern debate. As we will see, the best way for clarifying this concept is to study the history of the concept of emergence.

3. The Concept of Emergence[9]

In the modern literature we can find different meanings of "emergence". Sometimes emergence is explicated in such a way that emergence excludes reducibility.[10] Other people introduce emergence in a sense which is compatible with reducibility. Mario Bunge, for example, defines a property as emergent if it characterizes a system as a whole and is not possessed by the components of system.[11] Ernest Nagel calls a property of an object constituted by certain elements "emergent" if it is impossible to deduce this property even from the complete knowledge of the components, of their special relations, and of properties exhibited by other complexes formed by the same elements.[12]

The following short history of the concept of emergence will help to find criteria for choosing a fruitful variant and to preserve the crucial elements of the concept of emergence.

3.1 The composition of causes and the explanation of complex systems.

We start our history with J. S. Mill. He did not employ the word "emergence", but he introduced the fundamental distinction, which G. H. Lewes took up, when he used "emergent" for the first time (1874) in its technical meaning.[13]

In the 6th chapter of the III. book of his «System of Logic» John Stuart Mill deals with the «Composition of Causes». Regarding the joint effect of several agents or causes he distinguishes two ways in which the effects of causes acting separately from another are connected with the effect of their conjunct agency:

1. «...if we happen to know what would be the effect of each cause when acting separately from the other, we are often able to arrive deductively, or à priori, at a correct prediction of what will arise from their conjunct agency.»[14] This model of composition is characteristic for mechanical processes. Mill's persuasive example is the composition of forces and the resultant motion. In modern terms, he seems to think of linear superposition.[15]

2. In other areas of nature, such as those investigated by chemistry, the composition has special effects: «The chemical combination of two substances produces, as is well-known, a third substance with properties different from those of either of the two substances separately, or of both of them taken together. Not a trace of the properties of hydrogen or of oxygen is observable in

those of their compound, water. The taste of sugar of lead is not the sum of the tastes of its component elements ...»[16]

Later on, Lewes calls this case "emergent" and distinguishes it from the mechanical addition, which is "resultant".[17] Emergent properties are even more familiar in biology: «but the phenomena of life .... bear no analogy to any of the effects which would be produced by the action of the component substances considered as mere physical agents.»[18] In Mill's terms the joint effect of causes is different from the sum of their separate effects.[19]

John Stuart Mill touched nearly all the problems which were discussed in the following debate on emergent properties. The relation of separate effects and joined effect can be determined by a logical aspect (Is it possible to derive the joined effect in principle?) and by a epistemic aspect (Is it really possible to predict the joined effect?).[20]

A second field of discussion is demarcated by the examples which refer to secondary qualities (as taste, for example). They raise additional problems, but we shall not deal with this sophisticated debate.[21]

From a methodological point of view Mill's and Lewes' ideas of analyzing complex systems are incomplete. They focus on observable properties and underestimate the power of general laws. In the light of present theories of explanation the results of their analyses seem less convincing than they were when evaluated against the background of the philosophy of science of that time.[22] The following weaknesses are present in the proposals of Mill and Lewes: Concepts like "composition" or "sum" are too weak for developing a theory of emergence. The examples of resultant properties are mostly quantities which are simply added in the composition. The possibility of understanding the existence of new properties by using more complicated laws in describing the composition is generally underestimated. The formal properties of reduction are not analyzed.

3.2 The Emergence of Novelty[23]

The next period of the discussion on emergence started in the nineteen-twenties. The concept gave its name to a whole movement: "Emergent Evolutionism". The concept "emergence" even became well known in public, especially through the book *Emergent Evolution* (1923) by the biologist and philosopher C. Lloyd Morgan.

This movement was deeply influenced by Darwin's theory of evolution and strove for a philosophical understanding of the formation of new organisms (more generally, of new structures in nature).

The emergent evolutionists took a mediate position. On one hand they rejected metaphysical vitalism. The coming-into-being of the new structures should not be traced back to special forces acting from «outside». But a mechanistic view was also rejected. The "emergent evolutionist" did not believe in a program aiming to explain all new highly-organized systems on the basis of the knowledge of lower levels. (Of course, one could ask whether there really is any comfortable position between mechanism and vitalism!)

The emergent evolutionists were inclined just to accept the formation of new things as something «given» that cannot be explained. C. L. Morgan was ready to do so with an attitude of «natural piety».[24]

Some passages of C. L. Morgan will illustrate the ideas of emergent evolutionism. The main point is the emergence of novelty: «But the orderly sequence, historically viewed, appears to present, from time to time, something genuinely new. Under what I here call emergent evolution stress is laid on this incoming of the new.»[25]

The new structures constitute levels, which are characterized by special relations between the elements («relatedness»). The new relations, which cannot be found on the lower level, do not allow prediction of the emergent properties.[26]

The emergent evolutionists did not choose the strategy of applying the laws of the lower level to the new boundary conditions. In this way their antivitalistic attitude does not lead to reductionism. One reason for their antireductionism might be the fact that in the beginning of our century the prospect of finding physical explanations for complex processes in living beings (for example in embryology) seemed entirely hopeless. But there might have been additional philosophical reasons: «It may be that we have just to accept the newly given facts -all the facts as we find them- in the frankly agnostic attitude proper to science. Or it may be that in the acknowledgment of God an ultimate philosophical explanation, supplementary to scientific interpretation, is to be found. That will be the position I shall try to maintain.»[27]

Let us have a look into C. D. Broad's book *The Mind and its Place in Nature*.[28] He carefully analyzes and corrects Morgan's distinction between emergent and resultant properties. As he stresses, even in mechanics the knowledge of the properties of the separate constituents and of the form of composition does not suffice to predict (or in general, to know) the behaviour of the composed system.[29] We need an additional law of composition. This general law of composition can be gained much more easily in mechanics than in the typical examples of emergent properties. Broad gives the following explication of emergence: « ... the characteristic behaviour of the whole could not, even in theory, be deduced from the most complete knowledge of the behaviour of its components, taken separately or in other combinations, and of their proportions and arrangements in this whole. This alternative, ..., is what I understand by the "Theory of Emergence".»[30] Broad takes common salt (sodium chloride) as an example of emergence: «... so far as we know at present, the characteristic behaviour of Common Salt cannot be deduced from the most complete knowledge of the properties of Sodium in isolation; or of Chlorine in isolation; or of other compounds of Sodium such as Sodium Sulphate, and of other compounds of Chlorine, such as Silver Chloride.»[31] We must not forget that at this time quantum mechanics had just begun to develop a theory of chemical bonding. Broad does not give any general arguments against the possibility of explaining chemical compounds.[32]

Broad rejects mechanism for empirical reasons. He does not believe that in chemistry there are general laws governing the composition of constituents, laws which could be compared with the rule of superposition of forces.[33]

The theses of the emergent evolutionists have been much discussed, also among biologists engaged in thinking about the philosophical problems of their science. During the VI. International Congress for Philosophy (1926) a special session on emergence was organized. Hans Driesch was one of the participants.[34]

The problem of real novelty stressed the "ideological" component of the concept of emergence. The debate, which followed in the thirties, was mainly concerned with the task of finding a precise concept of reduction.

3.3 The logical analysis of reduction

Logical empiricism contributed to the clarification of "emergence" by analyzing the logical structure of physical theories and focussing on the formal aspects of reduction.[35]

In their famous 1948 paper C. G. Hempel and P. Oppenheim have a special section «On the Idea of Emergence».[36] Emergence becomes a relation between theories at a given time; there is not much interest in ontological questions:[37]

«The occurrence of a characteristic *W* in an object *w* is emergent relatively to a theory *T*, a part relation *Pt*, and a class *G* of attributes if that occurrence cannot be deduced by means of *T* from a characterization of the *Pt*-parts of *w* with respect to all the attributes in *G*.»

Emergence means that at a given time the properties of living organisms cannot be deduced from the physical and chemical laws which are known at that time. As the following discussion has shown, there are ontological problems going beyond questions of logic: Are the laws of physics sufficient to understand (at least in principle) the phenomena of life? Or are there any hints that physics is in some sense nomologically incomplete and life is directed by new laws which are not connected with the laws of physics? In any case, Hempel and Oppenheim taught an important lesson: Explicating "emergence" should be based on a modern concept of explanation.

In discussing such ontological questions we are not restricted to the logical and semantic relations of theories which happen to be formulated by scientists at a given time and are precise enough for formal investigations. The ontological problem has been illegitimately transformed into a question concerning only the state of knowledge at a given time. Of course, if we leave the solid ground of analytical methods, we have less reliable instruments for discussing emergence and reductionism.

The seminal paper «Unity of Science as a Working Hypothesis»[38] by Paul Oppenheim und Hilary Putnam mentioned in the second section of this paper was a move away from the peaceful realm of logic. They were interested not only in logical problems of theory reduction, but also in the assessment of the metaphysical program of a nomological unity of science. This new approach was influenced by the impressive success of molecular biology in deciphering the genetic code. Their paper was a fine example of the interaction of science and philosophy.

In his textbook *The Structure of Science* (1961), Ernest Nagel gives not only a classic explication of theory reduction, but he also deals broadly with emergence.[39]Already in a paper in 1949 he had developed the essential elements of his explication of theory reduction.[40] In describing the background of the problem of theory reduction he stressed that classical mechanics was assumed to be a comprehensive theory: «The success of that theory in explaining and bringing into systematic relation a large variety of phenomena was for a long time unprecedented; and the belief entertained by many eminent scientists and philosophers, sometimes supported by a priori arguments, that all the processes of nature would eventually fall within the scope of its principles was repeatedly confirmed by the absorption of several sectors of physics into mechanics.»[41]

He was well acquainted both with the metaphysical idea of unity of science and with the formal tools of theory reduction and took up the old problem of mechanism in biology. In his textbook E. Nagel starts with Broad's concept of emergence:[42] An object is constituted out of certain elements. It is assumed that we know all the properties the elements possess when they exist «in isolation» and that we also know all the properties exibited by other complexes formed out of these elements. If there is at least one property of the object which cannot be predicted by this knowledge, we call this property an "emergent property" and this object an "emergent object".

Contrary to Broad's writing, E. Nagel is careful in speaking about propositions and in dealing with logical relations. He gives the example of a mechanical clock and shows that even in physics there are emergent properties in the defined sense.[43] Nagel tries to separate the notion of emergence from its antimechanistic background. The existence of novelty is compatible with a reductionistic program: «The form in which the suggestion of emergent laws appears most plausible is that new types of behavior conforming to novel modes of dependence arise when hitherto nonexistent combinations and integrations of matter occur.»[44] Emergent properties are not required to be novel in a temporal sense.[45] Novelty is not the essential characteristic of emergence and the debate on emergence can be separated from the idea of cosmic evolution. Emergence does not exclude reductionism: Properties (or propositions on properties) which are emergent relative to one theory can be reduced by other theories.

4. Reductionism and Emergence: Some systematic lessons from history

In the following part of my paper I shall use the historical material just presented as a heuristic tool for an explication of the concept of emergence which is compatible with a reductionistic approach. I would like to discuss problems of "emergence" in the context of explaining complex systems and take the emerging of qualitatively new properties or specific laws as a criterion for the differences of levels.[46] For this reason, I do not recommend calling those properties emergent which are possessed by the whole, but not by its parts.[47] Following this definition nearly all mechanical systems (eg. clocks, and even chairs) would have emergent properties. This does not cover the type of problems discussed under the heading of "emergence" in the history.

My main idea is that emergent properties occur if the tools of explanation, which are sufficient for the parts of a whole, are not adequate for a real understanding of the composed system.[48]

But under what circumstances should a reductionist believe in this sort of emergence? One reason for doing so would be the failure of the methods and formulas physicists use for explaining simple systems (like a lone electron moving in an electric field) in understanding highly organized systems like a slipper animalcule. I think we should distinguish two senses of physics (or "physical"). In the first sense physics consists of the fundamental equations and all the consequences following from them through introducing the material content of the world (as boundary conditions). In this sense we use "physics" when speaking about "physicalism". In a second sense, we use "physics" as "physics of simple systems" which is the physics of one-particle systems but also the physics of multiple particles; here we can use special methods to reduce complexity (if we take the material content of the universe as an ideal gas, for example). In this sense we speak of «departments of physics» in our universities. Biology is clearly not reducible to physics in the second sense, because there are emergent properties, but it *might* be reducible to physics in the first sense. This mirrors in many respects the classical difference between ontological and methodological reductionism. So materialism - reduction in principal to physics in the first sense - can be held without maintaining that the *methods* of the physics of simple systems can be used to understand dung-beetles and sun-flowers.

The concept of emergence can be employed without following an emergentist philosophy, which K. Lambert and G. Brittan have nicely characterized: «There are phenomena which cannot be fitted into the general derivational pattern, or which cannot profitably be decomposed into their subatomic parts, or which resist the best efforts of scientists to provide a purely physical explanation for them. It is in these ways that the emergentist sets limits to science.»[49]

As I mentioned before, in the case of emergence the theory which is sufficient to understand the behaviour of the parts is *not* sufficient to understand the behaviour of the whole.

An example is oscillating electric circuit, which was discussed in this context by Konrad Lorenz.[50] In the explanation of this circuit we need more elements of Maxwell's theory and a new type of boundary conditions than in explaining the behaviour of a single coil or of a single capacitor. Thus, we have an emergent property (oscillations) and we can reduce this property to fundamental equations, as it is done in textbooks of physics.[51]

In a similar way the coming into existence of new organisms can be understood on the basis of laws which do not change in time: «Generally it can be said that a more classificatory or Aristotelian approach will be inclined to see novelty when the modern, explanatory approach sees continuity and likeness.»[52]

Separating emergence from anti-reductionism fits with the historical fact that authors with quite different backgrounds (such as Marx and Engels, Comte, Alexander, R. W. Sellars, Lovejoy, Broad) stressed the significance of emergence.[53]

I suppose that many emergent properties have their origin in our way of dealing with complex systems. We have to use fairly simple theories to explain even highly organized systems. So the notion of emergence has a pragmatic aspect. Moritz Schlick gave an pragmatic justification of holistic conceptualizations in a paper in 1935.[54] I think we should be consistent and face the pragmatic components of the notions of level and emergence. Fritz Rohrlich speaks about "cognitive emergence", arguing that our cognitive capacity is responsible for the pluralistic nature of our ontology. This pluralistic approach is compatible with the notion of unity of nature ("substantive monism").[55]

Philosophy of science should be more interested in working with complex systems. Strategies of explanations and the usual ways of solving equations, which are successful in simple systems, very often fail when applied to complex systems.

F. A. von Hayek, who has developed some interesting ideas for a theory of complex systems,[56] identifies the boundaries of the levels with the scope of successful application of methods of explaining simple systems: «Nonphysical phenomena are more complex because we call physical what can be described by relatively simple formulae.»[57] A reductionist can also account for employing different concepts and laws at different levels of Nature and explain why different strategies of explanation are successful. (Examples would be feedback circuits or descriptions which are confined to analyzing the function of various elements.)

The emergence of new pattern can be a result of the increase in the number of elements. Even if simple relations exist between the elements, the larger structure as a whole can possess features which are relatively independent of the particular values of the individual data: «Such "wholes", defined in terms of certain general properties of their structure, will constitute distinctive objects of explanation by a theory, even though such a theory may be merely a particular way of fitting together statements about the relations between the individual elements.»[58]

This is a consequence of the pragmatic aspects of scientific explanation. Higher-level explanations can be better than explanations using irrelevant atomistic details.[59] I have proposed calling those properties of complex systems "emergent" which cannot be explained by those parts of the fundamental theory which are sufficient to understand the behaviour of the isolated components. We have found various reasons why the theories sufficient for explaining the parts may not be useful in explaining the whole:

1. The explanation of whole systems requires additional elements of fundamental theory (laws of interaction, for example).

2. The complexity of the whole requires new methods of analyzing the phenomena (for example, concepts of information theory).

3. Special boundary conditions require strategies of explanation which are specific to the higher level. (An example: The usual statistical methods of multiple particle physics cannot be employed in biology because variations in the initial states are decisive.)

4. The macro-level exhibits constant structures, which are more useful in explanations than irrelevant microscopic details.

In the light of the same four reasons (and probably also from other ones) we can understand why ontological reductionism is compatible with the existence of emergence. Along this line of argument, we can do justice to a reductionistic program *and* to many of the arguments of emergentists we have uncovered in our historical survey.

Center for Philosophy and Foundations of Science - University of Giessen (Germany)

NOTES

1. I thank Clark Peddicord for improving my poor English.

2. Cf. *The Oxford English Dictionary*, Volume XIII (Oxford: Clarendon Press, 1989).

3. I have not found anybody using "reductionism" before C. Garnett, "Scientific Method and the Concept of Emergence," *The Journal of Philosophy* 39 (1942): 477 - 486. But Garnett does not seem to have invented the word "reductionism".

4. W. V. O. Quine, *From a Logical Point of View* (Cambridge, Mass.: Harvard UP, 1953), 20.

5. R. M. Hare, *The Language of Morals* (Oxford: Clarendon Press, 1952), 180.

6. M. Scriven, "Definitions, Explanations, and Theories," in: H. Feigl/M. Scriven/G. Maxwell (eds.), *Minnesota Studies in the Philosophy of Science*, Vol. II (Minneapolis: University of Minnesota Press, 1958), 123.

7. R. Oppenheim/H. Putnam, "Unity of Science as a Working Hypothesis, in: H. Feigl/M. Scriven/G. Maxwell (eds.), *Minnesota Studies in the Philosophy of Science*, Vol. II (Minneapolis: University of Minnesota Press, 1958), 3 - 36.

8. See for example Michael Ruse, "Sociobiology and Reductionism," in: Paul Hoyningen-Huene/ Franz Wuketits (eds.), *Reductionism and Systems Theory in the Life Sciences* (Dordrecht: Kluwer, 1989), 45 - 83.

9. After finishing the first version of this paper I learned that Achim Stephan (University of Göttingen) is preparing an extended history of the concept of emergence. In the meantime I had the privilege to read an unpublished manuscript of Dr. Stephan and to learn more on emergence in helpful discussions with him. David Blitz (Central Connecticut State University, New Britain) has written a dissertation entitled "Emergent Evolution: The Problem of Qualitative Novelty in the Evolution Process". Unfortunately I have not been able to consult his book, which will be published in near future. In any case, these authors will complete my preliminary remarks on the history of emergence in this paper.

10. J. J. C. Smart, *Philosophy and Scientific Realism* (London: Routledge & Kegan Paul, 1963), p. 50: "By an emergent law I mean one which relates to some complex entity, but which is in principle inexplicable in terms of the simple entities. It is important for me to deny the existence of such emergent laws, since it is a central thesis of this book that animals and men are very complicated mechanisms."

11. M. Bunge, "Emergence and the Mind," *Neuroscience* 2 (1977): 501 - 509 (cf. p. 502).

12. E. Nagel, *The Structure of Science* (New York: Hartcourt, Brace & World, 1961), 367.

13. George Henry Lewes, *Problems of Life and Mind* (London: Trübner & Co., 1874/1875), Vol. I p. 98, Vol. II p. 412.

14. J. S. Mill, *A System of Logic* (1843) (Toronto: University of Toronto Press, 1973), 370.

15. A hundred years later, Ernest Nagel disputed the thesis that this form of composition is the only one in mechanics. The latest developments in non-linear dynamics (for example in the theory of chaotic systems or in synergetics) confirm the objection raised by Nagel.

16. Mill (1843), 371.

17. Lewes (1874/75) does not yet exhibit the strong antimechanistic inclinations which are typical for the first two decades of our century. He seems more interested in the question of how it is possible to separate the effects of different causes if only the combined effect is known: "Thus, although each effect is the resultant of it components, the product of its factors, we cannot always trace the steps of the process, so as to see in the product the mode of operation of each factor. In this latter case, I propose to call the effect an emergent. It arises out of the combined agencies, but in a form which does not display the agents in action" (Vol. II, p. 412).

18. Mill (1843), 371.

19. Mill (1843), 373.

20. This difference has been stressed by C. W. Berenda, "On Emergence and Prediction," *Journal of Philosophy* 50 (1953): 269 - 274.

21. C. D. Broad took up this problem (cf. C. D. Broad, *The Mind and its Place in Nature* (London: Kegan Paul, 1925), p. 71. Cf. Wilfrid Sellars, "Philosophy and the Scientific Image of Man," in: W. Sellars, *Science, Perception and Reality* (London: Routledge & Kegan Paul, 1963), 1 -40. See also Arthur Pap, "The Concept of Absolute Emergence," *British Journal for Philosophy of Science* 2 (1952): 302 - 311, and Evelyn Pluhar, "Emergence and Reduction," *Studies in the History and Philosophy of Science* 9 (1978): 279 - 289.

22. Lewes (1875), Vol. II, "An explanation, to be valid, must be expressed in terms of phenomena already observed ..." (p. 7). See also Lewes (1874), Vol. I: "We are not to conclude the properties of elements from the properties of the groups they form; nor vice versa" (p. 97). Maurice Mandelbaum stresses the theory of explanation of the time: "...in Moleschott, Vogt, and Büchner ... , what might be denominated as the classic method of scientific explanation was assumed to be the proper method for science to follow: a scientific explanation consisted in analyzing the concrete events to be explained into a set of specific characteristics, attempting to establish constant relationships among various of these characteristics, and from these relationships deducing by ordinary logic

what consequences would follow in any given case." (Maurice Mandelbaum, *History, Man, & Reason* (Baltimore: Johns Hopkins Press, 1971), 26).
Cf. Mario Bunge, *Causality* (Cambridge (Mass.): Harvard University Press, 1959), chapter 8.2, especially p. 208 and p. 213. He shows that some doctrines of causality have the consequence that effects of a new type are excluded a priori.

23. This is the title of a book of Cowan Lloyd Morgan, *The Emergence of Novelty* (London: 1933).

24. Morgan, C. Lloyd, *Emergent Evolution* (London: Williams and Norgate, 1923), 16. This position was correctly characterized as crypto-vitalistic. See Zdzislaw Kochanski, "Kann Biologie zur Physiko-Chemie reduziert werden?" in: Bernulf Kanitscheider (ed.), *Materie-Leben-Geist* (Berlin: Duncker & Humblot, 1979), 82.

25. Morgan (1923), 1.

26. "Let there be three successive levels of natural events, A, B, and C. Let there be in B a kind of relation which is not present in A; and in C a kind of relation, not yet present in B or A. If then one lived and gained experience on the B-level, one could not predict the emergent characters of the C-level, because the relations, of which they are the expression, are not yet in being. Nor if one lived on the A-level could one predict the emergent character of b-events, because ex hypothesi, there are no such events as yet in existence. What, it is claimed, one cannot predict, then, is the emergent expression of some new kind of relatedness among pre-existent events. One could not foretell the emergent character of vital events from the fullest possible knowledge of physico-chemical events only, if life be an emergent chord and not merely due to the summation, however complex, of constituent a-notes. Such is the hypothesis accepted under emergent evolution." (Morgan (1923), 5 - 6).

27. Morgan (1923), 9.

28. C. D. Broad, *The Mind and its Place in Nature* (London: Kegan Paul, Trench, Trübner & Co., 1925).

29. Broad (1925), 63.

30. Broad (1925), 59.

31. Broad (1925), 59.

32. Broad (1925), 59.

33. Broad (1925), 66 ff.

34. Cf. E. S. Brightman (ed.), *Proc. of the 6th Int. Congress of Philosophy (1926)* (New York: Longmans, Green and Co., 1927), 1 - 46.

35. An early example is Gustav Bergmann, who stressed the role of initial conditions for the existence of novelty (Gustav Bergmann, "Holism, Historicism, and Emergence," *Philosophy of Science* 11 (1944): 209 - 221).

36. C. G. Hempel, P. Oppenheim, "Studies in the Logic of Explanation," *Philosophy of Science* 15 (1948): 135 - 175.

37. Hempel/Oppenheim (1944), 151.

38. In: H. Feigl/M. Scriven/G. Maxwell (eds.): *Minnesota Studies in the Philosophy of Science*, Vol. II (Minneapolis: University of Minnesota Press, 1958), 3 - 36.

39. Ernest Nagel, *The Structure of Science* (New York: Harcourt, Brace & World, 1961), p. 366 ff.

40. Ernest Nagel, "The Meaning of Reduction in the Natural Sciences," in: Robert C. Stauffer (ed.), *Science and Civilization* (Madison: University of Wisconsin Press, 1949), 99 - 135.

41. Nagel (1949), 99.

42. Nagel (1961), 367.

43. Nagel (1961), 373.

44. Nagel (1961), 380.

45. Nagel (1961), 375.

46. Cf. Harold Chapman Brown, "A Materialist's View of the Concept of Levels," *Journal of Philosophy* 23 (1926): 113 - 120.

47. Contra M. Bunge, "Emergence and the Mind," *Neuroscience* 2 (1977): 501 - 509 ("... emergent properties, or properties characterizing a system as a whole and which the system components do not have", p. 502).

48. Robert L. Klee, "Micro-Determinism and Concepts of Emergence," *Philosophy of Science* 51 (1984): 44 - 63, reports other approaches to "emergence". He also deals with the problem what could be called "emergent" (properties, laws, phenomena).

49. Karel Lambert, Gordon G. Brittan, *An Introduction to the Philosophy of Science*, 3rd edition (Atascadero, California: Ridgeview Publishing Company, 1987), 206.

50. Konrad Lorenz, *Die Rückseite des Spiegels* (München: Piper, 1973), 48.

51. Cf. Paul Hoyningen-Huene, "Zu Problemen des Reduktionismus der Biologie," *Philosophia Naturalis* 22 (1985): 271 - 286 (especially p. 283).

52. Gustav Bergmann, "Holism, Historicism, and Emergence," *Philosophy of Science* 11 (1944): 209 - 221 (especially p. 214).

53. Cf. Maurice Mandelbaum, *History, Man, & Reason* (Baltimore: Johns Hopkins Press, 1971), 380.

54. Moritz Schlick, "Über den Begriff der Ganzheit," *Wissenschaftlicher Jahresbericht der Philosophischen Gesellschaft an der Universität zu Wien* - Ortsgruppe Wien der Kant-Gesellschaft für das Vereinsjahr 1933/34 und 1934/35, pp. 23 - 37, reprinted in: Ernst Topitsch (ed.), *Logik der Sozialwissenschaften* (Köln: Kiepenheuer & Witsch, 1971), 213-224.

55. Fritz Rohrlich, "Pluralistic Ontology and Theory Reduction in the Physical Sciences," *British Journal for Philosophy of Science* 39 (1988): 295 - 312 (see p. 297).

56. F. A. Hayek, "The Theory of Complex Phenomena," in: Mario Bunge (ed.), *The Critical Approach to Science and Philosophy. Essays in Honor of Karl Popper* (New York: The Free Press of Glencoe, 1964), 332 - 349.

57. F. A. Hayek (1964), 336.

58. F. A. Hayek (1964), 336.

59. Cf. Hilary Putnam, "Philosophy and our mental life," in: *Mind, Language and Reality. Philosophical Papers*, Volume 2 (Cambridge: University Press, 1975), 296.

Erwin Engeler

THE TECHNICAL PROBLEM OF "FULL ABSTRACTNESS" AS A MODEL FOR AN ISSUE IN REDUCTIONISM

The goal of this paper is to use an *artificial context* of reduction of one world to another to illustrate two pheonomena which have so far not been described as types of *failure of reductionist methodology*.

The task before us is a modest one; we want to clarify certain aspects of epistemology, in particular the *issue of reduction*, and we want to use Leibnitz's paradigm of "calculemus" to do this. Hence, we seek a *formal* context where one could reasonably speak of reduction. The *intuitive examples* that are in the back of our formal considerations are the

$$\text{reduction pairs} \begin{cases} \text{"real" world /thepretical model} \\ \text{"deep" world / observable world} \end{cases}$$

Let me start with a very simple example. I want to show that *compositionality can* easily *be violated.* Assume that we have perfectly valid theories for the behaviour of two given objects A and B. Each of these is considered as a discrete input-output device: if I feed them a sequence of possible inputs, each will produce a series of outputs according to its inner workings. This

E. Agazzi (ed.), The Problem of Reductionism in Science, 91–99.

is the *observable* behaviour of A, resp. B, and a complete theory of that behaviour would define formally the set of pairs (u, v) of possible inputs and outputs, u and v codified by some sort of formal description, e.g. as words over a finite alphabet. Now, imagine a third device C which accepts either A or B as a component and devise an input-output theory of the more complex systems C plus A and of C plus B. The basis *expectation* is, that full knowledge of A and B and the connectivity of the composition would suffice to obtain full knowledge of the composite systems, i.e. that the input output relation of the composite system is determined by that of its components. This expectation is what would be called compositionality; it is a prejudice that could be ascribed to the methodology of reductionism: "reduce complex systems to their components, describe those fully and compose to a description of the full system". It fails, [2].

Namely, let A and B be *nondeterministic* devices which both read input streams of tokens "0" "1" and output corresponding streams of zeroes and ones. Device A operates according to the following options: either it first reads an input token and then outputs first "0" than "1" or it first outputs "0", then reads the next input token and after that outputs "0" again.

A option (a) : read; 0 ; 1
option (b) : 0; read; 0

The device B with which we will later replace A is described in a similar way as

B option (a) : read; 0 ; 1
option (b) : 0; read; 1
option (c) : 0; read; 0

The input-output behaviours of A and B can be described as the set of pairs of "words" (u,v), i.e. finite streams of tokens:

$\overline{S}$ (device) is the set of pairs (u,v) such that the device outputs v if the input stream is u . With this definition we observe that

$$\overline{S}\,(\mathrm{A}) = \{ <\varepsilon,\varepsilon>, <\varepsilon,0> \} \cup \{ <w, (01 \vee 00)^{n} >: \mathrm{length}\,(w) = n \neq 0\}$$

where ε is the empty word and $(01 \vee 00)^{n}$ is the set of words consisting of n parts, each being either "01" or "00". Similarly $\overline{S}$ (B) can be computed and turns out to equal $\overline{S}$ (A).

Now, let us compose either A or B with a third device C as follows. C simply copies each incoming token to its two outputs. The connectivity is as follows:

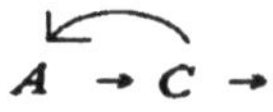

respectively,

Let $\overline{S}_{0}$.(device) now simply denote the output of the device without input. Then

$$\overline{S}_{o}(\mathrm{C\ plus\ A}) = 00(01 \vee 00)^{\infty}$$

$$\overline{S}_{o}(\mathrm{C\ plus\ B}) = (01 \vee 00)(01 \vee 00)^{\infty}$$

using the notation (of "regular languages") explained in an example above. These two behaviours are different and we have thus an example of non-compositionality.

Compositionality of the observed behaviour $\overline{S}$ fails, obviously because the deep behaviour also comprises things like reading and waiting. Let us introduce formal token □ ("box") for these discrete steps, then the deep behaviour S of A, respectively B, is a set of pairs (u, v) of words, now over the alphabet consisting of O, 1 and □ . It is obvious that S(A) is different from S(B) and then considerably less surprising that S(C plus A) is different from S(C plus B).

Let us now propose a framework in which the above two types of world (or behaviour-) descriptions, the "deep" description and the "observables" description are presented and compared in some generality. (The term "world" is used mainly for rhetorical emphasis.)

Deep World **W**		Observable World $\overline{\mathbf{W}}$
operational behaviour		input-output behaviour
$\mathbf{W} = \langle W, \cdot \rangle$		$\overline{\mathbf{W}} = \langle \overline{W}, \cdot \rangle$
$W = \{0,1,\ \}^*$	$\xrightarrow{f}$	$\overline{W} = \{0,1\}^*$
free monoid		free monoid
$A, B, C, \ldots, \delta$	$\xrightarrow{g}$	$\overline{A}, \overline{B} . \overline{C} \ldots, \delta$
formulas for description of operational behavior		formulas for description of input-output behavior
S operational semantics		$\overline{S}$ input-output semantics

$s(\delta) \subseteq W^k$ $\overline{S}(\delta) \subseteq \overline{W}^{k'}$

Reduction

(f, g)

Thus, a reduction pair is a pair of maps, the first one maps the operational (or "deep") behaviour of a device onto the (input output) observable (or "surface") behaviour. The second map is a map between the corresponding description languages, in our case the descriptions such as "option (a) : read ; 0 ; 1", etc.

What should be admitted as a *language* for the description of worlds, what should we accept as formal models of worlds in the first place? I hold that as "Worlds" we should accept any relational structure, but that we are best served for our intuition if we consider a world to be a phase-space, a product

$$W = \text{Space} \times \text{Time} \times \text{Data}.$$

A typical system is distributed (discretely or continuously) over space, and at each point in space which is occupied by a part of the system, there are present some data. These evolve along world-lines.

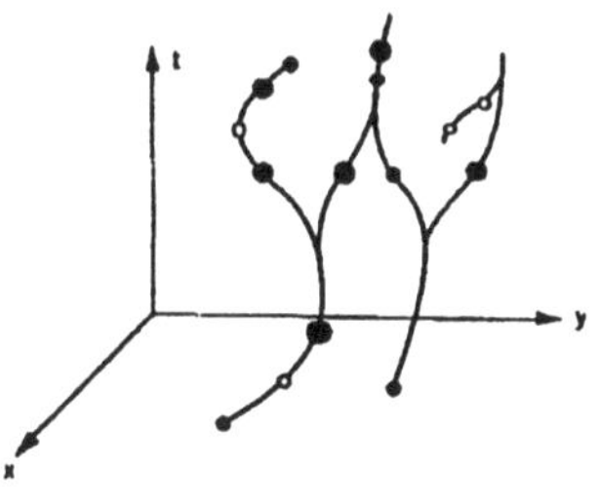

$W = \text{Space} \times \text{Time} \times \text{Data}$

B = system behaviour = set of worldlines $\subseteq$ W

W is W, furnished with additional struct•re: relations, operations on data in conjunction with space and time; creation etc. of processes.

B = $S(\delta)$, where S is the semantics of a description δ

If we do understand worlds as relational structures for the purpose of our analysis of the notion of reduction, the next question is the choice of languages for describing behaviours. Here, it is important to realize that descriptions using classical predicate calculi are restricting unnecessarily what can be said, e.g. of outcomes of experiments. A more comprehensive language would be one that describes subsets of the phasespace W by describing the experiments themselves, and these are programs (for the experiments). Technically, the corresponding description language is simply the quantifier-free predicate calculus, augmented by *recursive definitions* of predicates, starting with the basic, predicates and operations of the relational structure which we called "world". Thus, our original picture generalizes as follows:

$\mathbf{W}$		$\bar{\mathbf{W}}$
$W = S \times T \times D$	$\xrightarrow{f}$	$\bar{W} = \bar{S} \times \dot{T} \times \dot{D}$
relations, operations		relations, operations
a, b, c,...		$\bar{a}$, $\bar{b}$. $\bar{c}$ ...
$t_1, t_2, \ldots$		$\bar{t}_1, \bar{t}_2 \ldots$
programs	$\xrightarrow{g}$	programs
π		$\bar{\pi}$

e.g. logic programs

$a(t_1(X_1Y), t_{2y}))$	:-	$b(t_3(X), t_4(Y)), a(X,Y)$
$b(0,1)$	:-	
	...	

S semantics $\qquad\qquad$ $\overline{S}$ semantics

$S(\pi) \subseteq W$ $\qquad\qquad$ $\overline{S}(\overline{\pi})\ \overline{W}$

(f,g)
Reduction pair

Now that we have a formal set-up for the notion of reduction, we are in a position to clarify some of the intuitive notions that occur in the context of reduction and reductionism. And we can hope to exhibit a phenomena with which reductionist methodology could be confronted.

Propeties of reductions (f,g)

faithful: $w \in S(\pi) \quad \leftrightarrow \quad f(w) \in \overline{S}(\overline{\pi})$

modular: $S(\pi_1) = S(\pi_2) \quad \rightarrow \quad \forall\overline{\pi}: \overline{S}(\overline{\pi}/g(\pi_1)) = \overline{S}(\overline{\pi}/g(\pi_2))$

adequate $\overline{S}\ (g(\pi_1)) = \overline{S}\ (g(\pi_2)) \quad \rightarrow \quad \forall\pi: s(\pi/\pi_1) = S(\pi/\pi_2)$

fully abstract:

$$\forall\pi:\ \overline{S}\ (g(\pi_1)) = \overline{S}\ (g(\pi_2)) \leftrightarrow s\ (\pi/\pi_1) = s\ (\pi/\pi_2)$$

compositional: $\overline{S}\ (g(\pi_1)) = \overline{S}\ (g(\pi_2)) \rightarrow \forall \overline{\pi} : \overline{S}\,(\dot{\pi}/g(\pi_1)) = \overline{S}\,(\overline{\pi}/g(\pi_2))$

π/π_2 is the result of substituting π_2 in a fixed place in π.

In the title of my paper, I chose the concept "full abstractness" because I feel that this is the property which most extensively captures what we would call a good theoretical model of the world. - One of the irreparable failures of a universal reductionist methodology is that there are worlds for wich there is provably no fully abstract reduction pair. The example that I want to mention here originates again in theoretical computer science, [1, 3].

N		$\overline{\mathbf{N}}$
N	$\xrightarrow{identity}$	N
$\leq, t, ., 0, 1$		$\leq, t, ., 0, 1$
recursive programs		recursive programs
random assignment $x := ?$		

Let me finish with a disclaimer. I do not for a moment pretend that I addressed the whole issue of reductionism in the natural sciences, not even within mathematics.
But I do feel that the point of view here expressed also needs to be taken into consideration.

ETH - Zentrum, Zürich (Switzerland)

REFERENCES

(1) K.R. Apt & G.D. Plotkin, Countable nondeterminism and random assignment.
Journal of the ACM, 33 (1986).

(2) J.D. Brock & W.B. Ackerman, Scenarios: a model of non-deterministic computations .
Lecture Notes in Computer Science vol. 107. Springer Verlag, 1981.

(3) A. Stoughton, Fufly Abstract Models of Programming Languages .
Pitman Publ. Co., London, 1988.

Jules Vuillemin

A NEUTRAL REDUCTION: ANALYTICAL METHOD AND POSITIVISM

What is generally called *reduction* is a method allowing the elimination of a supposedly specific quality thanks to a definition that reconstructs that quality as a function of other qualities already admitted in the language of science.The paradigm of such a reduction is the kinetic definition of temperature as a function of pressure, volume and total number of atoms.

Such reductions have an immediate theoretical importance in physics. They bear witness to the explanatory character of a theory that annexes a branch of physics. The theory of heat is thus reduced to a compartment of mechanics.

It is remarkable that physics presents also another type of reduction that, far from intending to augment the explanatory power of a theory, proposes, quite on the contrary, to avoid theoretical decisions about the nature of things and to limit physical theory itself to a description of phenomena.

It is the analytical methods that are the instrument *par excellence* of these neutral reductions. We shall first describe them. Then we shall show their affinity with the positivist ideal and the permanent element that that ideal represents in physics.

E. Agazzi (ed.), The Problem of Reductionism in Science, 101–118.

I

THE ANALYTICAL METHOD

Those who have read Descartes will not have failed to notice that the idea of extension is subject to two different developments of a distinct nature in his work.

It is wrong to regard the *Geometry* as a treatise of analytical geometry, given that Descartes limits the object of his study to algebraic curves. The reasons he has, however, for excluding the so-called mechanical curves have to do rather with the narrowness of the constraints he imposes on the notion of analysis than with the notion of analysis itself. Indeed, to reject the sinusoid from the domain of analysis on the pretext that it has an infinite number of solutions is a prejudice. But what is essential in the Cartesian revolution lies elsewhere. It is that, unlike the Ancients, Descartes' geometry represents algebraic curves of any degree whatsoever in the plane. The Cartesian plane is therefore no longer a section of Euclidean space. The curves of which it treats may be «solid» or «hypersolid». The point of the coordinate axes is not to fix or to measure the plane figure projected by some real object, but to furnish a symbolic tableau of the law of correspondence between the independent variable x and the variable of the function y. Let us forget the restrictions imposed on the function. Then we do have the right to consider Descartes' *Geometry* to be the first treatise of analytical geometry.

Elsewhere, Descartes identifies physical body with its principal attribute, extension, which exhaustively defines the clear and distinct idea of body. Extension, thus identified with matter, can no longer be resolved in the framework of a functional correspondence. It is real, homogeneous, symmetric, endowed with three dimensions, infinitely extended and infinitely divisible. The identification of extension and body has as immediate consequence two theoretical theses in physics: there is no void, there are no atoms. If we add that the only modes of extension we are able to conceive of clearly and distinctly are motion and rest, a very precise physical doctrine results from Cartesian metaphysics: namely continuist mechanism, whose master concept is that of the vortex, and for which fluid mechanics will soon formulate the mathematical counterpart.

Advancing now to the end of the XVIIIth century, what do we see? The geometers have perfected, extended and systematized analytical geometry. Doubts have come to them about the implicit Cartesian conjecture according to which it is possible to solve all algebraic equations of any degree whatsoever by roots. But these very doubts give rise to work on auxiliary equations which illustrates the Cartesian rule of complete enumeration, when this last is properly understood. By contrast, Descartes' mechanistic physics - if we except the geometry of waves by which Huygens explains the laws of optics - has not produced any results. Huygens' hypothesis itself remains speculative, as is the contrary one of Newton's. What is more, theoretical quarrels abound in physics. Partisans of the caloric theory are opposed to the mechanists and no experimental decision seems possible. Even regarding just the surest part of science, Newton's rational mechanics, mustn't we regret that the physicists, instead of abiding by the theoretical prudence of Newton, by the *hypotheses non fingo* of the general Scholium, are letting themselves be led down the dogmatic path, are interpreting the general fact of universal gravitation in terms of attraction, are postulating the existence of another fundamental force, repulsion, and presume to explain the universe by the concourse of these two forces?

Laplace, Poisson are working in this spirit. On the one hand, they conceive of a mechanical system as a set of free material points subject to the real forces of attraction and repulsion, of which analysis must reveal the presence in the principleS of the most diverse of physical phenomena, such as chemical reactions or capillarity for instance. On the other hand, every physical phenomenon calls for a theoretical explanation. What is of fundamental importance here is not that Laplace should have opted for caloric, for luminous fluid, for electric fluid, but that he should have considered physics to be an explanation in which a hypothesis intervenes through its real content. Such an option continues the physical theory of Descartes in making a simple instrument of analytical geometry, admittedly important for seeing clearly, but destined solely to giving precision to the chosen hypothesis and putting it into an equation.

But one can assign another role to analysis. It is to the merit of Lagrange to have understood this in his *Mécanique Analytique* where he subjects mechanics to the same transformation Descartes had operated on geometry. To simplify, consider the motion of a material point M on an axis Ox. The position vector OM and velocity vector $V = d\frac{OM}{dt}$ have the same support Ox. Let us choose a perpendicular axis Oy upon which we shall put the measure of velocity v = Mm of the point M at each instant. Corresponding to the point M on Ox there is then the point m in the plane xOy of coordinates x = Om, y = Mm. At every instant the datum m informs us of the position and velocity and consequently of the state of motion, in the classical sense, of the material point. If M describes Ox according to a particular temporal law, m will describe a particular curve in the plane Oxy.

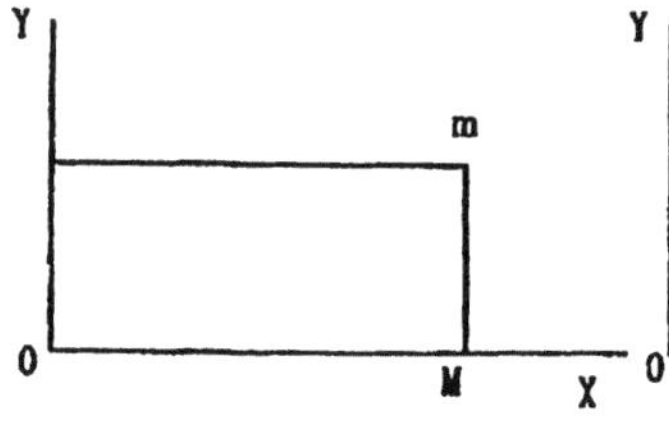

Uniform motion

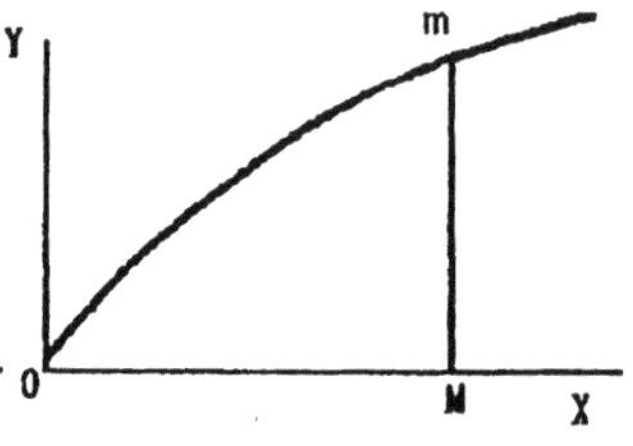

Uniformly accelerated motion
(v=at=m describes a parabola tangent to Oy at O)

The space Oxy is called a configuration space or extension in phase. A six-dimensional space-figuring the three spatial coordinates and the three components of velocity - will allow representing the state of motion of a particle. If a system is composed of N independent particles, its state will likewise be represented in a configuration space of 6 N dimensions. Such a space, as is shown even more by the dimensional heterogeneity of its coordinates (L and LT^{-1}) than by their number, is evidently not a mathematical copy of real physical space.

Let us consider the extension in phase of uniformly accelerated motion. We are given the initial conditions: $x(t_o) = v(t_o) = O$, $v = at$, $a =$ constant. From these givens we can immediately get : $x = \frac{1}{2} at^2$, and in eliminating t we get the Huygens relation : $v^2 = 2ax$, a purely kinematic equation which is put into dynamical dress by writing it as :

$$\frac{1}{2} mv^2 = max.$$

The equality is between a function of the velocity coordinate and a function of the position coordinate. It gives an elementary illustration of the two principles upon which Lagrange bases his analytical mechanics.

Imagine the line Ox concretized as an ideal rigid wire upon which a solid slides without friction but compelled to maintain its orientation. Thanks to the principle of virtual work, the constraints to which the motion of the solid is submitted will be expressed by setting the elementary work equal to O :

$$d\,\mathcal{L} = d(\frac{1}{2} mv^2) - d(max) = O$$

The first differential in the substraction above is none other than the product $(m \frac{dv}{dt})(vdt) = Fdx$, the work of a force due to actual motion. The second, $(ma)dx = dV/dx = fdx$, is due to the potential $V = max$. It may be regarded as a fictive or an inertial force. And the equation:

$$\vec{F} = \vec{f}$$

is an expression of d'Alembert's principle according to which there is equilibrium at every instant between the «real» forces (internal and external) and inertial forces, and thanks to which Dynamics may be formally reduced to Statics.

But let us return at present to the Huygens relation in its dynamical dress. Let us derive the function of the velocity coordinate with respect to that coordinate. Then let us take the derivative with respect to time of that derived function. This last quantity is the same as the derivative of the position function with respect to the variable of position, but in the contrary direction. Such is the Lagrangian equation that dominates the whole of analytical mechanics:

$$\mathcal{L} = T - V$$

$$\frac{d}{dt}\frac{\delta \mathcal{L}}{\delta v} - \frac{\delta \mathcal{L}}{\delta x} = 0$$

This equation (or the system of equations that generalizes it) owes its analytical importance to three circumstances: it conserves the same form when coordinate systems other than Cartesian ones are used; the position coordinates designate in general the independent parameters describing the instantaneous position of a body or set of bodies and the velocities then become generalized velocities, first derivative with respect to time ($\dot{q}_i$) of these generalized coordinates (q_i); in the general case the Lagrangian $\mathcal{L}$ is a function of generalized coordinates and velocities and can depend explicitly on time. And lastly, the Lagrangian equations are deducible from the principle of least action.

We shall examine further on the particular status that analytical mechanics reserves for the notion of force and in what sense we can say that it allows reducing dynamics to statics. It will suffice for the moment to comment on the program it announces in its very title. Descartes reduced geometry to analysis in studying a polynomial as a function of a variable. In the same way Lagrange reduced mechanics to analysis in expressing the state of a motion by a set of generalized coordinates and velocities, that is to say, of parameters in sufficient enough number for expressing the state of the motion in question at every instant, and in formulating the system of second order differential equations that analytically express the relations that exist among these different parameters.

Quite analogous remarks could be made regarding Fourier's *The Analytical Theory of Heat.* Fourier deduces the equation of the propagation of heat - known and generalized in mathematical physics under the name of equation of diffusion - from the three equations representing the general facts of thermology. The first of them is the differential equation of heat[1] conduction or Fourier's law:

$$h = -k \text{ grad } T,$$

(k being the thermic conduction capacity
of the substance considered)

which establishes the relation between heat flow and temperature distribution. The second is the differential equation of the conservation of heat :

$$-\frac{dq}{dt} = \text{div } h.$$

What it links is quantity of heat and flow of heat. The third is the differential equation establishing the relation between quantity of heat and temperature :

$$\frac{dq}{dt} = \rho c \frac{dT}{dt}$$

where ρ designates the density of the mass and c the specific heat of the substance. The equation of the propagation of heat is written:

$$\frac{dT}{dt} = \frac{k}{\rho c} \text{div grad } T.$$

We shall later examine the nature of the physical hypothesis that corresponds to the second equation, the one that postulates the conservation of heat, but we shall pass over the important developments of mathematical physics such as the theory of Fourier transformations and the vectorial calculus which were occasioned by Fourier's thermology. Here again Fourier reduces questions of thermology to questions of analysis. Experience suggests three quantities: flow of heat, quantity of heat and temperature, by means of three physical constants. The law of diffusion that results immediately from these three fixed hypotheses by way of these same three constants is a relation of proportionality between the rate of temperature change over time and the result of the so-called Laplacian spatial operator applied to the same temperature.

II

POSITIVISMS

It is the analytical theories that nurtured positivism. Without them it would not have seen the light of day nor have developed as it has.

One can distinguish three forms of positivism: the properly analytical positivism, linguistic positivism and «holist» positivism, incarnated respectively by Comte and Mach, by Carnap, by Duhem and Quine. We shall examine its postulates only insofar as they have a bearing on the nature of hypotheses in physics.

Comte drew his inspiration directly and expressly from Lagrange and from Fourier.

For him, «the eminently philosophical revolution brought about by Lagrange in his admirable treatise of *Analytical Mechanics* represents a radical reconstruction of the entire system of rational mechanics where, with statics treated according to a sufficiently general primitive law of equilibrium and with dynamics brought back to statics, the whole of science was able to acquire a character of unity henceforth irrevocable»[2].

Just what reductions does this unification involve? Comte has been taken to task for having made Lagrange say that mechanics had no need of the notion of force[3]. But the appeal to the principe of virtual work shows the inanity of this reproach. When Comte declares that «forces, in mechanics, are nothing but the motions produced or tending to be produced», he is not eliminating force but is rather restoring it to its Newtonian meaning. It is the cause of finite or differential motion, or rather, it is identical to it, as can be seen from the expression of work in the principle of virtual work. What Comte is dismissing and criticizing when he protests against the substitution of the metaphysical word *attraction* for the positive word *gravitation* is the representation, under the term *force*, of an entity that would not have as sole function the describing of a state of motion. He is doing no more here than explaining Lagrange. If it were argued further that the reduction of Dynamics to Statics by means of the principle of d'Alembert is purely formal, this too would be picking a false quarrel with positivism, since the motion produced by the force takes place over time and since Comte proposes calling rational mechanics *phoronomy*, which designates kinematics.[4]

As has aptly been said, Lagrange's formula is in fact «completely independent of Mechanics, it is a result of the kinematics of mass»[5]. And the reduction to the kinematics of mass has precisely as aim the analysis of the notion of force. If Lagrange «invokes motion» - and the same could be said as regards Comte - «it is not in order to explain force, but only to set into correspondence with that idea, which transcends geometry, a numerical symbol capable of figuring in formulas»[6]. This indeed constitutes an extension of the method used by Descartes in his *Geometry*. Only here, the algebraic calculus has been replaced by vectorial calculus in assuming, as point of departure, a few «general facts» corresponding to the equilibrium and the constraints and fairly directly suggested by experience.

As for the *Analytical Theory of Heat*, it brings out even more clearly the method of analysis in the study of the phenomena of propagation. The three laws it postulates - and the three empirical coefficients entering into the expression of these three laws - are rather directly inferred from suitably idealized observation. It is mistaken to read Comte and Fourier from the logical positivist's point of view as regards the meaning of observation. Comte often expresses himself in contemptuous terms with regard to empiricist philosophy, which constructs its chimeras on immediate observation - that is to say that it admits sensible and even affective qualities as ultimate givens. It is not surprising then that metaphysical prejudices and would-be causal explanations should make good bed-fellows with that subjective experience. The idealization needed for transforming this primitive given into a scientific one is precisely the correlative of analysis. It supposes that the observation already relies upon, an instrument of measure making it possible to correlate a qualitative diversity with a quantitative one susceptible of treatment by the calculus. Analytical positivism, then, does nothing but systematize the method of Galileo. Instead of inquiring into the nature of things it aims at substituting for that nature an order accessible to the science of number.

The objections an empiricist would make against the idealizations in thermology then would simply be misinterpretations. Fourier reasons about a solid to avoid the phenomena of convection. He neglects radiation at a distance and friction. He gives the solid an infinite dimension to avoid the discontinuities of temperature, or disregards the dissipation of heat at the external surface of the ideal solid under consideration. The different approximations called into play in the use of any experimental apparatus take these difficulties into account. But what we were talking about here are experimental hypotheses that are irreducible to empirical givens, but are nevertheless legitimate in science. Fourier and Comte, however, go further still. They insist on the elimination from thermology of all hypotheses about the nature of heat. There lies the positivist rule: «knowledge of the mathematical laws to which the effects of heat are subject is independent of all hypothesis; it requires only attentive examination of the principal facts attested by common observation and confirmed by precise experiments».[7]

But then again, as has been often pointed out, the equation of propagation supposes the equation of continuity. This latter immediately evokes hydrodynamics. It seems then that in spite of their repeated denials Fourier and Comte made an essential borrowing from the theory of caloric - a fluid analogous to water. Bachelard has formulated the objection particularly clearly: «in taking away elasticity, Fourier confers on the agent impenetrability with regard to itself. It is this impenetrability, moreover, that will entail, in the very important theoretical cases where there is question of a manner of operation, the fundamental equation of the problem. All that goes in must go out, that is the condition of permanence. It is this materialist ground, amazingly solid, that will support the entire edifice. There is but this one physical idea in such a great work, it is the principle that determines all the equations».[8]

This analysis is not diriment however. As Mach points out, the concept of temperature is a concept of level. The real analogy that supports thermology, he says, is between Fourier's law on the one hand, and the interpretation of the Galilean law of falling bodies in terms of Newtonian potential on the other. In this latter case, the *accelerations* of equalization are determined by differences of distance; in the former the *velocities* of equalization are determined by differences of temperature. Once the criticism is put back in empiriocriticism, that is to say, once analysis is put back in experience, Mach is the direct and faithful heir of analytical positivism. It is only that the concept of energy and the law of conservation that defines it have been extended: energetism unites mechanics and thermology and replaces them.

It postulates a phenomenological unity of nature obeying the «principle of becoming».

«According to this principle, any change taking place in nature could come down to an exchange of energy and, from this point of view, all forms of energy would be considered to be autonomous and altogether equivalent. To each form of energy would correspond a factor of intensity proper to it: for gravitation this factor would be the height, for heat it would be the temperature. It is the difference in factors of intensity that would determine the manner in which the phenomena take place over time».[9] General phenomenology then, by which energetism is defined, allows the grouping of phenomena which are intrinsically the most heterogeneous under a selfsame mathematical analogy, that of stationary current.[10] The properly mechanical hypotheses having to do with force then lose their *raison d'être*. Thermology annexes mechanics.

Planck has shown the superficial character of the analogy between the law of falling bodies and the law of the propagation of heat. The former has to do with individual and reversible movement, the latter with irreversible statistical dynamics. The energeticians themselves, moreover, felt a need to reform the notion of analogy. The hydrodynamic analogy that is necessary for reaching Fourier's law, far from being immediately justifiable by a law of development relative to the existence of potentials, is itself one hypothesis among others which together constitute Energetics. No more than its companion hypotheses, is this particular one verifiable or falsifiable on its own. They face the tribunal of experience collectively. The Duhem thesis has been seen to be at work in his *Treatise on Energetics and General Thermodynamics*.[11] This synthetic energetism becomes understandable once it is seen that on refusing to explain the hydrodynamic analogy by recourse to the kinetic, that is to say atomistic, hypothesis, one is obliged to formulate adequate analytical hypotheses. Their *sui generis* character disallows their being reduced to simple experimental inductions.

I have left logical positivism to the side. This last divides truths into two well distinct classes. There are the analytic truths on the one hand, the synthetic ones on the other. This separation by the axe is, as Quine says, untenable. In the first place, mathematical analysis is itself creative and cannot be reduced to the tautologies of logic. Once the inviability of the logicist program is recognized, it is more appropriate to establish the separation at the point where quantification is over classes. And everyone agrees on the synthetic - though *a priori* - character of the truths of set theory. Then again, we are unable to isolate the empirical synthesis from analysis, at least in so far as it is required by scientific experimentation. Quine is right in rejecting the analytic-synthetic dichotomy as it was construed by the logical positivists.

Logical positivism was an episode. Is that a reason though, for thinking that the Duhem-Quine thesis should be accepted? Duhem rightly insists on the difference of nature between mathematical and physical truths, a difference sufficiently established when we compare the cumulative history of mathematical demonstrations and theorems to the apparently highly erratic history of physical theories. Quine, by contrast, in the undifferentiated unity of his logico-mathematico-physical hypotheses, is not obliged to distinguish acceptable hypotheses of the analytical type from metaphysical ones such as atomism. In other words, there is nothing that binds Quine to phenomenalism, a feature of the thesis proper to Duhem, and his doctrine of the collective verification of hypotheses is compatible with a realist interpretation of science.

The insensitivity of Quine's thesis to the distinction between two sorts of hypotheses assures him an apparent advantage in guaranteeing to his methodology a unity and simplicity that are absent from the methodology with which his own indifference to history has imprudently associated him. But this advantage is only apparent. From the time of Osiander, from Ptolemy, no, from even as early on as Plato's famous *«sózein ta phainómena»* the history of physical theories has been divided between those that can be called *positivist* in the broad sense of the term, which either limit admissible hypotheses to inductive ones or perhaps give them a little more leeway but in forbidding in any case that they should speculate as to the nature of things, and those theories that could be called *explanatory* and that claim to get hold of the real causes of the phenomena. There is a reason to be found in the history of science that should incline us here to the complex dualistic methodology of Duhem: it is that this methodology, extraneous in principle to the history of mathematics, admits of raising the question of a conflict that was and still remains the main spring of the history of physics.

Let us return then to Duhem's abstract method, in asking ourselves what light it can throw on the conflict of hypotheses. Time and again Duhem presents a hierarchy of hypotheses entailing a growing ontological commitment. Let us limit ourselves to describing the two extreme terms.

Aristotelianism, and the analytical theories and in particular energetism following upon it, refuse to reduce the plurality of qualities - ground of the plurality of the categories and of the sciences - to a supposed unity. To this plurality there is a corresponding plurality of independent hypotheses. It may happen nevertheless that these analytical hypotheses, necessary for rendering the phenomena truely physical, that is to say accessible to the calculus, reveal a remarkable identity of form. It is noticed for example that electrostatic potential, the flow of heat and the diffusion of neutrons all follow the same equation. If we follow Aristotle in speaking of *analogy* in cases of the occurrence of a same law in domains having to do with different categories, these identical equations prove the existence of physical analogies. But these analogies, far from revealing an underlying substantial unity of nature that would justify reducing one class of phenomena to another more fundamental one, simply reflect the identity of idealizations required for experimentation in the different cases. They disappear when we fine-tune the margin of approximation required for one of the phenomena for example.[12] As Comte has already pointed out, physical analogy, which is expressed precisely in the differential laws characteristic of the analytical method, results from the simplicity assumptions that prevail when we compose complex phenomena out of elementary ones. It is like when in the development of a function in a series we retain only the first term because its contribution is such that it allows our abstracting from the other terms as a first approximation. There is a ground for this process in things and that is why we can consider our classification to be a natural one. It is chimerical to wish to extend our pretentions and to demand of experiment that it decide between two explanatory hypotheses that may happen to come into confrontation over a single phenomenon. In a given state of science, experiment depends on setting up an equation relative to a stated margin of approximation. To believe in the possibility of crucial experiments is to surreptitiously transform this stated approximation into a fixed and definitive one.

At the other end of the scale is Cartesianism: not the Cartesianism of the *Geometry* but the Cartesianism of physics, the one that reduces matter to extension. Such a hypothesis is explanatory in that it reduces the diversity of phenomena to the unity of an essential attribute. Here it is the irreducibility of the diversity of qualities that is denied. Because of this, the diverse hypotheses that constitute a unitary theory must form an organic whole, in contrast to the simple analogies required by Aristotelianism. Theoretical conflicts are not concerned with the mathematical expression of phenomena but rather with decisions about the nature of things. Although these decisions in question are always expressed relatively to a stated margin of approximation and subsequent refinements of that margin may entail theoretical revisions, the decisions themselves pertain to the logic of confirmation and of falsification through experiment. In this sense crucial experiments must exist.

The Michelson-Morley experiment can be regarded as a crucial experiment with respect to the unification of space and time in relativistic chronogeometry. Classical mechanics implicitly postulated that the vector sum of two velocities was obtained by the law of vector addition. It thereby accorded a meaning to a physical velocity of a modulus increasing beyond all limit. Measurements of black body radiation furnish another example of what a crucial experiment is. Thermodynamics had the advantage here of verifying Stefan's law: the pressure of radiation varies in proportion to the fourth power of the temperature. On the other hand, it was not known how to determine the coefficients in play. Classical kinematics held out promise of providing this determination. Unfortunately, its calculations led to absurd physical predictions and violated Stefan's Law (Rayleigh's Law). In formulating a law in accordance with the experimental results, Planck falsified one of the particular hypotheses of classical kinematics, concerning the continuity of the emission and the absorption of energy.

CONCLUSION

If positivism proposes describing phenomena abstractly and limiting itself to that description, Duhem's thesis has the merit of making the logic of positivism explicit. It offers the advantage of liberating analytical hypotheses from the over narrow inductive constraints put upon them by Lagrange, Fourier and Comte in the first positivism. Since science gives up speculating about causes to simply represent phenomena by the calculus, it is easy to imagine what this representation will inevitably contain of conventional. Why demand that the laws of Cartesian optics should be derived exclusively in the language of Newtonian corpuscular theory or in the wave theory language of Huygens? There are translations between the two, and it is these translations that are representative. Beyond that begins metaphysics, which is legitimate in its own right but where experience ceases to be decisive. Rejection of the crucial experiment is thus a consequence of physical phenomenalism. To demand of experience that it decide between beams and waves is to abandon physics and to confuse analytical reduction, which is legitimate but only more complex than induction, with metaphysical reduction, which is illegitimate and goes beyond our faculty of knowing.

Duhem's thesis is thus a dogmatic one. It is not an historical conjecture. And it is the dogma that constitutes its weakness. Neither relativity theory nor the introduction of quanta into physics find an echo in Duhem's doctrine, blind as it was to the double revolution taking place in science at the beginning of the century. The ideas of simultaneity and continuity in the emission and absorption of energy had so much apparent evidence for them, that they had remained implicit in earlier theories, positivist and analytical ones included. How can we escape seeing that to challenge them is to raise a doubt about a physical belief? How can we avoid interpreting the contrary hypotheses as explanatory ones, all the more in that they surprise the expectations of common sense? How can we resist the ultimate indisputability of experience? It is true that the differential laws discovered by the analytical method hardly allowed distinguishing between nature and our contrivances of classification. But we truly do run up against the universal constants c and h. The burden of proof goes over to the other side then, and it is hard to see how to interpret these numbers if not as the admission that nature is telling us some of her secrets.

Collège de France

NOTES

1. Joseph Fourier obtains it in chapter II, section V, paragraph 128.

2. Auguste Comte, *Cours de philosophie positive*, Paris, 1892-1894, 6 volumes, lesson 16, p. 250.

3. On the criticisms addressed against Comte see Anastasios Brenner, *Duhem, science, réalité et apparence*, Paris, Vrin, 1990, pp. 116 *sqq*.

4. Henri Poincaré, *La science et l'hypothèse*, Paris, Flammarion, 1968, p. 118, summarizes the doctrine of the *Mécanique Analytique* as follows: «What is force? Louis Lagrange answers that it is a cause that produces the motion of a body, or that tends to reproduce it».

5. Joseph Perès, *Mécanique générale*, Paris, Masson, p. 192. Louis de Broglie (*La physique nouvelle et les quanta*, Paris, Flammarion, 1937, pp. 30-31) insists on the kinematical character of Lagrange's theory whose equations do not figure in quantum mechanics where «the dynamical notions retain their value while the kinematical ones become less clear: the quantity of motion, which in the classical ideas appears to be a magnitude derived from velocity, appears in quantum mechanics rather as a fundamental and autonomous magnitude, independent of the concept of velocity, a concept whose meaning no longer seems to be well defined in every case»

6. Pierre Duhem, quoted by A. Brenner, *op.cit.*, pp. 118-119. Duhem thinks he is criticizing Comte. In fact he has misread him. He thinks further that Lagrange took his inspiration from Leibniz who attributed an absolute and metaphysical meaning to momentum, a scalar quantity. Another error. The inspiration comes from Huygens, the Cartesian. For Lagrange there is no question at all of maintaining that force is a primary, irreducible and unanalyzable quality. He holds it rather to be reducible, through the operations of the calculus, to a measurable quantity, potential energy and kinetic energy.

7. Joseph Fourier, *Théorie analytique de la chaleur*, chapter I, section II, paragraph 22, p. 16.

8. Gaston Bachelard, *Etude sur l'évolution d'un problème de physique. La propagation thermique dans les solides*, Paris, Vrin, 1928, p. 36.

9. Max Planck, *Initiation à la physique*, Paris, 1941, p. 56, p. 258; Planck describes the theory of Ernst Mach, *Prinzipien der Wärmelehre*, 4. Aufl., Berlin, Leipzig, 1923, pp. 55-56, pp. 341-342.

10. Ernst Mach, *op.cit.*, pp. 402-403.

11. Paris, Gauthiers-Villars, 1911, two volumes.

12. R. Feynmann, R. Leighton, M. Sands, *The Feynman Lectures on physics*, 3 volumes, Reading Mass., Addison Wesley, 1970, 11-12.

Paul Weingartner

REDUCTIONISM AND REDUCTION IN LOGIC AND IN MATHEMATICS

The paper is divided into three parts. First I shall talk about reductionism and reduction in general. In the second chapter a historical example of reductionism will be analyzed: the principles of rationalism of the 17th century. In the third chapter different kinds of reductionisms and reductions in modern logic and mathematics will be discussed.

1. I shall begin with three points on reduction stressed by Popper: (1) A successful reduction is a great success and therefore scientists have to be reductionists. (2) Hardly any major reduction in science has ever been completely successful. (3) We can learn an immense amount even from unsuccessful reductions.[1]

1.1 There are a lot of examples of (1) in mathematics. Some of them are more simple like the reduction of rational fractions to ordered pairs of numbers and of these to unordered pairs of unordered pairs. A little bit more sophisticated is the following example:
Assume we are given a system of equations of the following form:[2]

$$a^1_1x^1 + a^1_2x^2 + \ldots + a^1_nx^n = y^1$$

$$a^2_1x^1 + a^2_2x^2 + \ldots + a^2_nx^n = y^2$$

$$\ldots\ldots\ldots\ldots\ldots\ldots\ldots\ldots\ldots\ldots$$

$$\ldots\ldots\ldots\ldots\ldots\ldots\ldots\ldots\ldots\ldots$$

$$a^n_1x^1 + a^n_2x^2 + \ldots + a^n_nx^n = y^n$$

E. Agazzi (ed.), The Problem of Reductionism in Science, 119–147.

Wanted: all solutions $x = (x^1, x^2, \ldots, x^n)$. If considered independently, there is no a priori knowledge about the existence of such solutions nor about their uniqueness. By some suitable reduction one can get some more transparency: One introduces the vector-space $\mathbb{R}^n$, the elements of which are the n-tupels $x = (x^1, x^2, \ldots, x^n)$ of real numbers x^i and considers the linear mapping $A: \mathbb{R}^n \to \mathbb{R}^n$ the matrix of which is determined by the parameters a^i_k. Now the system of equations can be written in the simple form $Ax = y$ and it suggests to study the linear mapping A.

We have a reduction from the complex to the simple (a whole system of equations is reduced to one equation) but at the same time this reduction reveals a richer algebraic structure (linear mappings can be added and multiplied) and offers new connections: For example if the reduced equation has the form $(E - A)x = y$, where E is the mapping, then a solution of the following kind could be found (guided by geometrical series):

$$x = \frac{1}{E-A} y = \sum_{k=0}^{\infty} A^k y$$

Provided certain conditions, this leads to correct results. The main point is: The series $\sum_{k=0}^{\infty} A^k y$ on the right side has to converge. That means one has to introduce a suitable notion of convergence (a "topology") into $\mathbb{R}^n$ and to require (in analogy to the geometrical series) that A is respectably small. The solution is rewarding for numerical purposes, too.

From this example it can be seen that the reduction is one from the complex to the simple only in one respect, in another respect (richer algebraic structure, new connections) it is the other way round. At the same time it is going from the more applied form to the more general form in respect to the generalisations and abstract auxiliary constructions used even if only as a rewarding round about.

This example already shows that a reduction can have opposite properties if the aspect or point of reference is different: Thus it can lead from the complex to the simple in one aspect and from the simple to the complex according to another. Therefore one cannot expect that reductions would go only in the direction indicated by the arrow in the list below. But it may be claimed there is at least one important aspect in every reduction which goes that way.

1.2 List of properties of reduction

	→	Examples:
unknown	known	17th cent. Rationalism
less known	better known	Hilbert Program
defined	primitive	17th cent. Rationalism, Logicism
non-transparent	transparent	Logicism, Intuitionism
non-intuitive	intuitive	17th cent.
learned	inborn	Rationalism
complex	simple	Logicism
non-pure	pure	Hilbert-Program
more presuppositions	less presuppositions	Intuitionism,Hilbert-Program, Logicism
stronger presuppositions	weaker propositions	Intuitionism
applied form	general form	Logicism, Relevance
non-unified	unified	Logicism
redundant	non-redundant	Relevance

theories	axiomatizable theories	17th cent. Rationalism
theories	complete theories	17th cent. Rationalism
theories	number theoretic model of theories	Leibniz
proof methods	mechanical-combinatorial proof methods	Leibniz, Hilbert Program

2. *Reductionism of 17th century Rationalism*

The Rationalism of 17th century, especially that of Descartes and Leibniz is a good example of a reductionist program, not only for philosophy but also for the sciences in general. I shall describe this program in its most important principles and it will be plain that they are reduction-principles of different kind. For several reasons I have chosen the method of stating explicitely the respective principles rather than giving a description of the philosophies of Descartes and Leibniz which is anyway not the topics of this paper. In another paper[3] I have done this in much more detail. There I tried to show that these principles have also a good support in the writings of both authors and can be supported by the many references which I have given there.

2.1 Principles of Scientific Knowledge

2.11 Proposition p is scientifically known iff it is reducible to a first principle. p is reducible to a first principle iff p is itself a first principle or it is derivable from a first principle - or can be traced back to a first principle by logical principles of deduction and definition.[4]

2.12 First principles can be known by intuition[5]

2.13 There are first principles which are known by intuition: "Thus each individual can perceive by intellectual intuition that he exists, that he thinks, that a triangle is bounded by three lines only, a sphere by a single surface, and so on."[6]

"The primitive truths of reason are those which I call by the general name of identicals ... Those which are affirmative are such as the following: everything is what it is, and in as many examples as we maydesire, *A* is *A*, *B* is *B*"[7]

"As for the primitive truths of fact, they are immediate internal experiences of an immediacy of feeling. And it is here that the first truth of the Cartesians or of St. Augustinus occurs: 'I think, therefore I am, i.e. I am a thing which thinks' ..."[8]

2.14 Everything known by intuition is known clare et distincte[9]

2.15 Descartes: Everything known clare et distincte is true.[10] Leibniz:Everything known clare et distincte and known to be possible (consistent) is true.[11]

From the principles stated so far it follows that for Descartes every first principle is true. For Leibniz it follows that every first principle is true provided the ideas put together in the principle are possible (consistent).

2.16 Every proposition derived from true premisses is true.

The principle is intrinsic in Descartes' description of the method of deduction.[12] According to Leibniz principles of deduction belong certainly to the truths of reason. Secondly they are hypothetical truths not existential truths. The above principle is intrinsic in many passages and deduction principles he states explicitely. His examples are modes of syllogism but he also points out that there are valid principles of deduction not included in syllogism and gives principles of the logic of relations as examples.[13] The principle of the reductio ad absurdum is one other main principle of deduction.[14]

2.2 Principles of Scientific Analysis

2.21 Term (concept, idea) *t* is scientifically analyzable iff it is reducible to primitive terms. *t* is reducible to primitive terms iff *t* is itself a primitive term or it can be traced back to primitive terms by a chain of definitions:[15]

> "Analysis is as follows: Let any given term be resolved into its formal parts, that is, let it be defined. Then let these parts be resolved into their own parts, or let definitions be given of the terms of the (first) definition, until (one reaches) simple parts or indefinable terms."[16]

> "When definition pushes analysis until it reaches primitive notions, without presupposing anything whose possibility requires an *a priori* proof, the definition is perfect or essential."[17]

As Rescher puts it "Leibnizian analysis is a logical process of a very rudimentary sort, based on the inferential procedures of *definitional replacement* and *determination* of *predicational containment* through explicit use of logical processes of inference."[18]

2.22 All scientific terms (concepts, ideas) which are primitive and derived from the mind are innate.

2.23 There are primitive terms (concepts, ideas) which come from the mind itself.

Descartes uses mostly epistemical examples like knowledge, doubt, ignorance ... etc. whereas Leibniz uses either mathematical ones like square, triangle, circle or metaphysical ones like being, possibility, equality.[19]

2.3 Principles of Deductive Systems

2.31 Descartes: Every deductive system is finitely axiomatizable.[20]

2.32 Leibniz: Deductive systems for logic, mathematics and metaphysics are finitely axiomatizable. Deductive systems for physics and jurisprudence (including ethics) are not finitely axiomatizable.[21]

Leibniz on the other hand made an important distinction here: In respect to the *vérités de raison* such axiom systems can be found; in this case the *vérités innées* are the axioms from which the other truths of reason can be derived. Also in respect to those contingent truths which are about ones own inner states (i.e. that I think, that I will) including ones own existence a finite axiomatization by man is possible. However this does not hold of the other contingent truths. Although they are incorporated in a deductive system which is not finitely axiomatizable it is not the case that man can find such an axiom system. The reason is that man would need infinitely many steps to trace back these contingent truths to simple axioms. Only God knows the axiom-system with one act of thinking (without thinking in finitely or infinitely many steps).

2.33 Leibniz: All finitely analytic propositions are necessarily true. A proposition p is finitely analytic iff p is reducible to an "Identical" in a finite number of steps. The reduction to an identical (like $A = A$ or AB is A) is carried through by a chain of definitions:

> "Quoniam vera propositio est quae identica est, vel ex identicis potest demonstrari adhibitis definitionibus."[22]

This is not - as some scholars think[23]- Leibniz's principle of non-contradiction because Leibniz explicitly calls the following principle the principle of non-contradiction:

> "... the principle of contradiction or identity, that is, that a proposition cannot be true and false at the same time"[24]
> "The principle of contradiction is in general: A proposition is either true or false."[25]

Concerning the second citation Leibniz speaks of two consequences, which I have discussed elsewhere.[26]

According to my interpretation which I have elaborated in more detail in my (IMS), the principle 2.33 is a soundness principle, i.e., a principle of the consistency of axiomatic systems: Every proposition which is derivable (in a finite number of steps) in the system is (necessarily) true.

If this interpretation is correct Leibniz at least used - in addition to different forms of the traditional principle of contradiction including the reductio ad absurdum principle - a principle for the consistency of axiomatic systems or a soundness-principle in the form of 2.33. That Leibniz indeed used such a principle is supported by many passages in which he describes the method of analysis.

By this method first of all the axioms can be shown to be true: Since they are identical propositions of a simple form (first principles) they are evident insights of the human mind. In most cases if not in all they are vérités innées according to him. Then the theorems have to be shown to be true. This is possible since every step (of the finitely many) can be watched and the transformations with the help of definitions and inference rules[27] can be checked. But since the principle 2.16 - that the (correct) logical consequences of true propositions are true - is accepted one can prove that the theorems (which have been derived in the axiomatic system) must be true too:

> "Jtabque cujuscunque veritatis reddi potest ratio, connexio enim praedicati cum subjecto aut per se patet, ut in identicis, aut explicanda est, quod fit resulutione terminorum. Atque hoc unicum summumque est veritatis criterium, in abstractis scilicet neque ab experimento pendentibus, ut sit vel identica vel ad identicas revocabilis."[28]

According to the last passage the principle 2.33 can also be expressed in this way: If a proposition (which can be incorporated into a deductive system) has a (sufficient) reason, i.e. if it has a proof by which it can be traced back to identical propositions by using definitions then it is true (and consequently also consistent).

2.34 Leibniz: Every true proposition is finitely or infinitely analytic.

Or more explicitly:

2.341 Every true proposition of logic, mathematics or metaphysics is finitely analytic (i.e. is reducible to "identicals" in a finite number of steps).

2.342 Every true proposition of physics (natural science) and of jurisprudence (including ethics) is infinitely analytic (i.e. is reducible to "identicals" in an infinite number of steps).

There are numerous passages to support 2.34 (and 2.341 and 2.342). Two very general ones are these:

> "that nothing is without a reason, or that every truth has its a priori proof, drawn from the notion of the terms, although it is not always in our power to make this analysis."[29]
>
> "Generally, every true proposition (which is not identical or true *per se*) can be proved *a priori* by the help of axioms, or propositions true *per se*, and by the help of definitions or ideas. ... It is certain, therefore, that all truths, even the most contingent, have an *a priori* proof, or some reason why they are rather than are not. And this is itself what people commonly say, that nothing happens without a cause, or that nothing is without a reason."[30]

From the citations and the other references given it is clear that Leibniz had a logical form or version of the ontological principle of sufficient reason, i.e., the principle that every truth has an a priori proof with either finitely or infinitely many steps. In my (IMS) I have defended the thesis that this logical version of the principle of sufficient reason is a completeness principle for deductive systems or for scientific truths which are ordered axiomatically or "more geometrico". In other words

2.341 and 2.342 say that logic, mathematics and metaphysics (built up "more geometrico") are complete systems, whereas physics and jurisprudence are only complete for someone who can grasp an infinite number of proof-steps (i.e. for God only). That is also the latter are complete in principle but it goes beyond the capacity of humans to carry through effectively this proof.

2.4 The reduction of Characteristica Universalis and of Ars Combinatoria.

In general:

All scientific languages are reducible to a mathematical language. All scientific arguments are reducible to mathematical calculations.

In more detail:

2.41 Every primitive term (predicate) can be represented by a characteristic basic number (or by a characteristic ordered pair of basic numbers). Every compound term (predicate) can be represented by a characteristic number (ordered pairs of numbers) which is the result of applying a certain mathematical function to basic numbers (ordered pairs of basic numbers).[31]

Leibniz first used basic numbers to write down syllogistic modes in this mathematical language. But this didn't work properly. Therefore he used later ordered pairs of numbers (where the first is positive, the second negative) for the predicates in syllogistic modes (subject-term, predicate-term, middle-term). The intensional relation of inclusion: "the predicate is contained in the subject" (like: living thing is contained in animal, i.e., the properties of living things are a subset of the properties of animals)[32] he interpreted mathematically as division between ordered pairs of numbers:

2.42 The ordered pairs of numbers have to be chosen in such a way that if the predicate is contained in the subject the number-pair representing the predicate term has to be contained mathematically in the number-pair representing the subject term.

2.43 The proposition q follows from the premises $p_1 \dots p_n$ iff there is a certain mathematical function such that the ordered sequence of numbers representing q is equal to the result of applying this function to the ordered sequence of numbers which represent the premises $p_1 \dots p_n$.

Leibniz has tried quite a number of different possibilities for a mapping of different categorical propositions to characteristic numbers. It seems that only Leibniz's last proposal was successful.[33] By this proposal Leibniz found a mathematical decision method for syllogistics.[34] Leibniz thought that such a mathematical calculus can be extended beyond syllogistics:

2.44 Provided the scientific system is built up in a mathematical language described in 2.41 - 2.43 then there is an ars inveniendi, i.e. a mechanical and combinatorial method with the help of which one can find every (or almost every) truth:

> "In Philosophy, I have found a means of accomplishing in all sciences what Des Cartes and others have done in Arithmetic and Geometry by Algebra and Analysis, by the *Ars Combinatoria.* ... By this all composite notions in the whole world are reduced to a few simple ones as their Alphabet; and by the combination of such an alphabet a way is made of finding, in time, by an ordered method, all things with their theorems and whatever is possible to investigate concerning them."[35]

> "Methodus inveniendi perfecta, si praevidere possimus, imo demonstrare antequam rem aggrediamur, nos ea via ad exitum perventuros; perfecta magis illa quae nullis utitur theorematis apus alios demonstratis, vel problematis ab aliis solutis."[36]

Hermes interprets - I think correctly - Leibniz's ars inveniendi as recursive enumerability.[37] The claim is then: there is a recursive function the values of which are all true propositions (the set of true propositions would be a recursively enumerable set). We know of course - since Gödel - that this is not the case. Hermes thinks that the claim of the ars inveniendi does not mean decidability (i.e. recursive enumerability of the set of true and of the set of false propositions). But Leibniz's ars judicandi seems to be to him a good candidate for that.

The program "more geometrico" and of the mathematization of all sciences of Decartes and Leibniz was one of the greatest programs of reduction ever proposed. A lot of reduction-pairs are involved in this program. Thus the principles of 2.1 can be characterized by the arrows non-transparent to transparent, non-unified to unified, less known to better known, those of 2.2 by defined to primitive, complexe to simple, those of 2.3 additionally by a reduction to a consistent, complete, finitely axiomatizable set of propositions or to a set which has a number-theoretic model. Meanwhile we know several limitations of this program. But finding out the places where the program doesn't go through revealed a lot of new knowledge.

3. Logicism

3.1 Logicism is an example of reduction which was as a whole unsuccessful but still one can admit that it was partially successful. Generally speaking one can distinguish five possible cases for the disciplines logic (L) and mathematics (M): they coincide, M is a subset of L, L is a subset of M, they partially coincide, they are separated.

Now Russell had one of the most extreme views in his Principles of Mathematics:

> "The fact that all Mathematics is Symbolic Logic is one of the greatest discoveries of our age; and when this fact has been established, the remainder of the principles of mathematics consists in the analysis of Symbolic Logic itself."[38]

Frege has not, it seems, claimed that the whole of mathematics can be included in logic. But Frege certainly has claimed that arithmetic can be reduced to logic:

> "It has often been said that arithmetic is only a more highly developed logic; but that remains disputable as long as the proofs contain steps that are not performed according to acknowledged logical laws, but seem to rest on intuitive knowledge. Only when these are resolved into simple logical steps can we be sure that arithmetic is founded solely upon logic. ... A dispute can only arise, so far as I can see, because of my fundamental law about 'ranges of values,' which perhaps has not yet been specifically expressed by logicians, though it is in their minds when, e.g., they speak of extensions of concepts. I hold that it is purely logical. In any case the place is indicated where the decision has to be made."[39]

3.2 The three main theses of Logicism are these:

(1) All specific mathematical terms are definable on the basis of the logical vocabulary.

(2) All mathematical arguments can be formulated in the framework of logic

(3) All specific mathematical theorems follow from axioms of logic and logical inference rules.

ad (3): To begin with the third, Russell did not accept it any more in his *Introduction to Mathematical Philosophy* (i.e. in 1919):

> "We may take the axiom of infinity as an example of a proposition which, though it can be enunciated in logical terms, cannot be asserted by logic to be true."[40]

There is a further reason for not accepting (3) given in the mentioned book: To make existence assumptions is not the business of logic:

> "There does not seem any logical necessity why there should be even one individual. ... The primitive propositions in Principia Mathematica are such as to allow the inference that at least one individual exists. But I now view this as a

> defect in logical purity. ... If that be so, no principle of logic can assert "existence" except under hypothesis."[41]

The assumption that there is at least one individual in the universe of discourse is an assumption of Standard Logic (Standard Predicate Calculus of First Order). The assumption is not explicit but implicit in the use of variables, though there are some theorems which reveal it more
explicitely, e.g. proposition *24.52 in *Principia Mathematica.*

Thus logic is too weak in order to derive all mathematical theorems from it in a threefold way:

(a) Existence assumptions are not principles of pure logic. But mathematics assumes the existence of numbers or sets ... etc.

(b) Even Standard Logic with its existence assumption of at least one individual in the universe of discourse is much too weak since in any mathematical theory there is at least one variable with a domain of reference of much higher cardinality than one (at least that of the natural numbers).

(c) The axiom of infinity does not follow from any axiom of Logic.

ad (1): The arithmetization of mathematics began in the 19th century. A first step was to reduce various kinds of numbers (up to complex numbers) to natural numbers by using explicit definitions. In a second step Frege went on to reduce natural numbers to cardinal numbers of certain concepts in the sense of: the cardinal number of concept F=df the extension of the concept equinumerous with the concept F, where 'equinumerous' is defined thus: the concept F is equinumerous with the concept G =df there exists a one-one-correlation between objects falling under F and objects falling under G.

So far the reduction is not complete because with this strategy only natural numbers (integers) have been reduced to logic but not sets of natural numbers. This last step Russell tried to take in *Principia Mathematica* with the help of the socalled ramified type theory. This theory was developed in order to avoid the antinomies and is therefore not directly connected or dependent on the program of logicism. However as Gödel says the reduction is not complete here either:

"So the question whether (or to what extent) the theory of integers can be obtained on the basis of the ramified hierarchy must be considered as unsolved at the present time. It is to be noted, however, that, even in case this question sould have a positive answer, this would be of no value for the problem whether arithmetic follows from logic if propositional funcitons of order n are defined (as in the second edition of *Principia*) to be certain finite (though arbitrarily complex) combinations (of quantifiers, propositional connectives, etc.), because then the notion of finiteness has to be presupposed, which fact is concealed only by taking such complicated notions as "propositional function of order n" in an unanalyzed form as primitive terms of the formalism and giving their definition only in ordinary language.[42]

Further obstacles are that the theorems of mathematics when transformed into the language of logic (by reducing mathematical concepts to logical ones) do not only presuppose additional existence assumptions but seem also not to have an adequate form in respect to negation. Thus the transformation of 7 + 5 = 12 into the language of logic is:

$$[\mathrm{E}!_7 x)Gx \wedge (\mathrm{E}!_5 x)\,Hx \wedge \neg(\exists u)(Gu \wedge Hu)] \rightarrow (\mathrm{E}!_{12}x)\,(Gu \vee Hu)$$

Here it is presupposed that there are enough entities in the universe of discourse, not just at least one which Standard Logic presupposes. Otherwise one can derive contradictions. If we try to negate the above sentence we run into difficulties in respect to the interpretation of G and H. If they are treated as free variables then there are existential quantifiers in the negation. And this would mean according to Frege that $7 + 5 = 12$ be analytic but $7 + 5 \neq 13$ synthetic.[43]

Independendty of that the above "reduction" shows another strange feature: it is a "reduction" from the simple to the complex and may therefore be doubted to be a reduction at all.

Logicism aimed at a lot of properties of reduction which were listed at the beginning. Thus logic was understood as the better known (or better understandable), more simple, more transparent, more pure, more primitive, more unified, more general theory which makes less presuppositions.

This understanding of logic was based on some picture about logic which was too simple and to a great extent in the spirit of Leibniz's program of 17th century rationalism (see above). The essential limitations of this program were discovered already mainly between the two wars by Gödel, Tarski and others. But it was also based on a picture of the relation between logic and mathematics which was too simple: Logic is the more general and more universal theory. But experience with the program of logicism and mathematical practice showed that the general logical structures are not so rewarding in mathematics[44] as structures which are, though general, still mathematically specific in a certain sense like groups, categories, sheaves, etc.

On the one hand, the kind of generalization introduced by set theory led to a lot of new interesting and important results for both logic and mathematics.[45] But such further reductions like those to write simple number theoretic equations in the language of Symbolic Logic are certainly an over-exaggeration in the wrong direction which are moreover inadequate to a considerable extent.

4. Existence in Mathematics

Existence in mathematics, is it reducible to consistency? Or to put it more accurately: The existence of an object with certain properties in mathematics, is it reducible to the mutual consistency of these properties? This is a question which was dealt with not only in philosophy of mathematics but also in the foundations of mathematics in connection with two other concepts: with satisfiability and with constructibility. With satisfiability because for example Gödel's completeness theorem just states the coincidence of consistency with satisfiability. And with constructibility because almost all variants of mathematical intuitionism agree in the thesis that existence in mathematics coincides with constructibility.

But looking first at mathematical practice we see that never existence assumptions are drawn from proofs of consistency. Only the other way round is the one which is supported by mathematical practice: assumptions of consistency are drawn from establishing models, i.e. from proofing existence. Or to put it in other words:

> "The usual proofs of consistency are proofs of the satisfiability of conditions, more accurately: of the satisfiability of conditions concerning ideal objects."[46]

This observation of mathematical practice shows in fact the place where the reduction is not complete: the coincidence of consistency and satisfiability is valid only for a rather restricted class of systems, i.e. for those which are complete. In all other cases we do have only the relation (implication instead of equivalence) which mathematical practice supports: to try to establish models or to proof satisfiability and so to infer consistency.

The underlying property for the reduction of existence in mathematics to consistency is certainly that from non-unified to unified may be also from non-transparent to transparent. As a matter of fact the reduction in the sense of unification is only possible in those systems which are unified in another sense, in the sense of being complete.

5. Intuitionism

Are the restrictions imposed by Intuitionism on classical mathematics a kind of reduction?

5.1 Let us ask what is the most important restriction in respect to which almost all different variants of Intuitionism agree. It is the thesis that existence in mathematics coincides with constructibility. According to this assumption a statement $(\exists x)Fx$ can only be established if one provides a construction of a natural number k and a proof of $F(k)$. On the other hand, in order to prove $\neg(\exists x)Fx$, for each individual n a proof of $\neg F(n)$ is required. That is, the translation of $(\exists x)Fx \vee \neg(\exists x)Fx$ is this: Either there is a construction of a natural number k and a proof of $F(k)$

or there is a uniform proof that shows the falsity of $F(n)$ for each n. But clearly it might be that neither of the above alternatives is true.

If the range of 'x' in Fx is finite then the validity of $(\exists x)Fx \vee \neg(\exists x)Fx$ has never been questioned because then $(\forall x)\neg Fx$ is nothing but a conjunction of a finite number of decidable statements. But if the range of 'x' in $(\forall x)\neg Fx$ is infinite one cannot give for each individual n a proof of $\neg F(n)$ because the process of persuing an infinite domain can never be finished.

Because of the validity *of* $(\exists x)(Fx) \vee (\forall x)(\neg Fx)$ in classical mathematics we get an indirect existence proof of $(\exists x)(Fx)$ whenever we succeed in deriving a contradiction from $(\forall x)(\neg Fx)$ (or from $\neg(\exists x)(Fx))$ and so establish $\neg[\forall x(\neg Fx)]$ (or: $\neg[\neg(\exists x)(Fx)])$. Intuitionistically, however, an indirect proof of that sort proves only the negative fact that it is not the case that there is a uniform proof that shows the falsity of Fx for every x. And from this one can only conclude the absurdity of the absurdity of $(\exists x)(Fx)$, i.e. $\neg\neg(\exists x)(Fx)$ or, in other words, the non-contradictoriness of $(\exists x)(Fx)$.

Thus for Intuitionism a contradiction does not yield a method for constructing a suitable object: Identifying mathematical existence with consistency means degrading mathematics to a game.

5.2 What areas of mathematics are affected by intuitionistic restrictions?

Arithmetic and Algebra are not so much affected, but Analysis is. Thus for example the property of connectedness - i.e. that for arbitrary natural numbers a, b it holds (classically) that $a = b \vee a > b \vee b > a$ does not hold in general intuitionistically. And this affects of course many classical proofs in analysis. As a consequence a very general difficulty is the impossibility to base a proof on the alternative that the discriminant is either 0 or a positive integer or a negative integer. Therefore much more complicated proofs are required. Such proofs have been found and discovering them and their methods has been also rewarding. Other restrictions concern for example the theorem of Bolzano-Weierstrass and theorems about Dedekind-cuts. These theorems are either false or meaningless. But they are very powerful in classical mathematics.

5.3 How then could one describe the kind of reductionism caused by Intuitionism? If one looks through the list which I have given at the beginning, then the result is not straightforward: Certainly Intuitionism is not a reduction from the complex to the simple. On the contrary, the proofs are usually more complex compared to that in classical mathematics. It is also not - at least not uncontroversially - a reduction from the unknown to the known or from the less transparent to the more transparent or from the non-pure to the pure. Moreover it is not a reduction in the style of 17th-century Rationalism, i.e. in the direction of axiomatizable and complete theories or in the direction of mechanical and combinatorial proof-methods.

Intuitionists themselves certainly claim that it is a reduction of the less intuitive and less clear assumptions to the more (or only) intuitive and clear ones. But what is more intuitive and clear: To base mathematical considerations from the very beginning on the notion of continuity or to build the continuum up with the help of set theory from discreteness, from the denumerable set of integers? It is difficult to make a decision here.

6. Hilbert's Programme

As Kreisel points out[47] there are at least two important meanings of what is called 'Hilbert's Programme'. In the wide sense the programme aimed at establishing "the adequacy of deductive formalism for the representation of intuitive branches of mathematics".[48] In the narrow sense - which is the one usually referred to - it aimed at proving the consistency of the usual formalizations of mathematics by finitist means. Especially the program in the narrow sense was supposed to have the following steps: First the mathematical theory (arithmetic, analysis) should be formalised as an axiomatic system. As a second step it should be shown by simple and evident means that the application of the rules of proof to the axioms could never lead to a contradiction. This metastatement should therefore be argued by finitary means, i.e. with finitist proof.

6.1 In what sense is Hilbert's Programme a reduction? Reduction can be seen from two angles: First as leading to an aim for itself. In this sense the known, the simple, the pure, the intuitively transparent to which the unknown, complex, non-pure and non-intuitive (non-transparent) is reduced is a value in itself, a last purpose. On the other hand the reduction can lead to a means for another goal. In this sense 17th century rationalism wanted to reduce the learned to the inborn because it was assumed that the latter has the maximal security.

Hilbert's Programme seems to be mainly a reduction in the second sense: The aims are stated above in respect to the wider and narrower understanding of the Hilbert Programme. The aim of saving mathematics from the antinomies and paradoxes was included in both aims stated above even if more concretly in the second one. So was the "final solution" of the problem of foundations.

But the means which Hilbert wanted to use for reaching these goals was thought to be peculiar and necessary not just sufficient: The reduction of all mathematical reasoning to finitist resoning.[49] It was thought to be peculiar because Hilbert seemed to have had a similar view as it can be found in 17th century rationalism: that it is possible to reduce all truths to simple and evident truths[50] (cf. ch. 2.11-2.23). And these truths are then the basis, the foundation. They can be found in all sciences[51] and so there is not only a chance for unifying the different branches of mathematics but also many other sciences.

6.2 First of all it should be mentioned that the Hilbert programme in its original sense can be carried through for certain parts of mathematics, for example for arithmetic without induction based on the classical predicate calculus. On the other hand it is not correct that finitist reasoning leads to a fully adequate understanding of the concepts involved. For example, although the first order theory of the addition of natural numbers has a decision method it has also a model containing "non-standard" integers. Thus not only incompleteness-theorems are limitations but also completeness and decidability results are limitations in another sense. This was seen relatively early by Skolem.[52]

6.3 Second, as Kreisel points out, the novelty of Hilbert's work is his proof theory or the idea that proof itself is to be made the object of mathematics.[53] This requires of course an understanding of proof and especially a detailled analysis of "finitist proof" or "constructive proof". Such an analysis was not worked out to such a detail that Hilbert's programme could be called precise or that the understanding of these notions would be sufficiently adequate. For instance a partial axiomatization of the above notions of proof has never been given within Hilbert's programme.

6.4 In general - as has been mentioned already above - some of the most important presuppositions underlying Hilbert's programme seem to be basic assumptions in 17th century rationalism expecially in Leibniz:
(1) That a purely arithmetical truth has a purely arithmetical proof. According to Leibniz a proof which is reducible to simple identity statements about natural numbers.
(2) That mathematical truths can be based on finitist proofs. In Leibniz's sense: that mathematical truths are "finitely analytic", i.e. are reducible to simple identity statements by a finitist proof which according to Leibniz is a proof which has a finite number of steps and in which only chains of definitions and transparent identity principles are used and which terminates with or reduces to something like 1 = 1. The metastatement that the proof-method terminates is for Leibniz the evident insight (not particularly mentioned) and for Hilbert the "only" purely arithmetical statement needed in the case of analysis.
(3) That the whole of mathematics has itself a logico-mathematical structure. According to Leibniz it can be represented by a system which is finitely axiomatizable and consists of necessary truths which are finitely analytic.
(4) To mention finally a difference: The mathematical activity according to Hilbert is almost exclusively the "ars judicandi"[54] (the method of proving theorems) whereas for Leibniz it consists in both "ars inveniendi"(the method of finding the first and most simple axioms)[55]and "ars judicandi".

The presuppositions (1) - (4) were refuted by Gödel's incompleteness theorems. And so were the underlying reductionist thesis of all mathematics to simple arithmetico combinatorial proofs, to finitist proofs, to first simple and evident insights.

7. Reduction and Substitution

Three directions of reduction mentioned at the beginning (ch. 1.2) were that of the complex to the simple, of the non-pure to the pure and of the applied form to the general form. Now the process of substitution in valid formulas (say of Propositional or Predicate Logic) seems to go in the opposite direction: from the simple to the complex, from the pure to the non-pure, and from the general form to the applied form.

Thus $(r \wedge \neg s) \vee q \rightarrow (r \wedge \neg s) \vee q$ as a substitution instance of $p \rightarrow p$ is more complex, non-pure and applied than the latter. Thus the question what is the most reduced form in this sense is the question what is the ultimate starting point for substitution in respect to valid formulas. The answer is: It is that form f of a valid formula which is maximally general in respect to substitution. That is, there is no more general form f^* such that f can be obtained from f^* exclusively by applying substitution and f is not identical with f^*. Thus $p \rightarrow p$ is such a general form and so is $p \rightarrow (p \vee q)$ or $p \rightarrow (q \rightarrow p)$ or modus ponens (as an implicational formula) and many other valid formulas. Many of them are usually taken as axioms just because of this property: to be maximally general in respect to substitution. I have used this condition elsewhere as one part of a relevance criterion and for a definition of "scientific proposition".[56]

Question: Is the reduction class, i.e., the class of all those valid formulas of classical propositional logic which are maximally general in respect to substitution as strong as the set of theorems of classical propositional calculus? Or: Do we loose anything by reducing all valid formulas of propositional calculus to those which are valid and maximally general in respect to substitution? The answer to the latter question is: no, we didn't loose anything, we can get everything more complex by applying substitution. Thus this reduction class is in some sense the gist or kernel of the system (in this case propositional logic) in respect to substitution.

8. Reduction and Relevance

The reductions of the redundant to the non-redundant, of the complex to the simple and of the non-pure to the pure seems also to hold for relevance. The general idea underlying the kind of relevance I have in mind here is this: A part of a valid formula is redundant (irrelevant) iff one can replace this part by an arbitrary part (formula) salva validitate of the whole formula. This general idea underlies different relevance criteria which have been proposed by Körner, Cleave, Schurz and myself.[57]

A precise definition of the most general criterion – since it applies to formulas in general, not only to consequences of formulas – is this: The valid formula α (of Propositional Logic or First Order Predicate Logic) is relevant iff

(1) It is not the case that a propositional variable (or predicate) in α is replacable on some of its occurences by an arbitrary propositional variable (or arbitrary predicate of same arity) in some positive proper part(s) of α salva validitate of α.

(2) No proper part of α is a tautology or a contradiction.

(3) α is maximally general in respect to substitution.

Then β is a relevant consequence of a iff $\alpha \rightarrow \beta$ is (valid and) relevant (in the above sense).

Question: Is the set of all relevant consequences of A as strong as A? The question has a positive answer for propositional logic. It was proved with a weaker criterion of relevance which is implied by the above one.[58] For Predicate Logic we think that the answer is positive too, though it seems difficult to carry out the proof without using some principles of second order logic. The reduction set obtained by the above relevance criterion is important and rewarding for the following reasons:

(1) It can be shown that most of the paradoxes in the Theory of Confirmation, Explanation, Law Statement, Disposition Predicates, can be avoided by applying this relevance criterion.

(2) It can be shown that most of the paradoxes in Deontic Logic, Epistemic Logic, and Value Theory can be thrown out, too.

(3) It can be shown that the problem of truthlikeness (verisimilitude) in the sense of Popper's original idea can be resolved and the paradoxes which came up in connection with it can be avoided.[59]

Salzburg University (Austria)

NOTES

1.Cf. K. R. Popper, (OUN), p. 132 f.

2.The example is taken from Cigler-Reichel (Top), p. 1 f.

3.P. Weingartner, (IMS).

4.Descartes (RD) 3, 4, Leibniz (GP) 7, p. 194 and 296.

5.Descartes (RD) 3; Leibniz (GP) 5, p. 343, (NE) 4, 2, 1.

6.Descartes (RD) 3.

7.Leibniz (GP) 5, p. 343; (NE) 4, 2, 1.

8.Leibniz (GP) 5, p. 347; (NE) 4, 2, 1.

9.Descartes (RD) 3; Leibniz (GP) 4, p. 423.

10.Descartes (RD) 3; (DM) 2; (DM) 4, 3.

11.Leibniz (GP) 4, p. 422; (GP) 1, p. 384, (GP) 7, p. 194.

12.Descartes (RD) 3, (RD) 12, 23.

13.Leibniz (GP) 5, p. 461.

14.Leibniz (GP) 6, p. 612.

15.Descartes (RD) 4, (DM) 2, (RD) 12, 13.

16.Leibniz (GP) 4, p. 64, (De Arte Combinatoria).

17.Leibniz (GP) 4, p. 450.

18.Rescher (LIP), p.22.

19.Cf. Descartes (RD) 12, 14; Leibniz (NE) 1, 1, 18; 1, 3, 3; 2, 1, 2; (GP) 5, p. 66, 79.

20.(RD) 3, 8; (RD) 5.

21.Leibniz (GP) 7, p. 57; p. 25; p. 194 f., (GP) 6, p. 612.

22.Leibniz (GP) 7, p. 195.

23.Cf. Rescher (PhL), p. 26; Copleston (HPh) Vol. IV, p. 277.

24.Leibniz (GP), 7, p. 355.

25.Leibniz (GP) 5, p. 343; (NE) 4, 2, 1.

26.Cf. my (IMS), p. 161.

27.Cf. Leibniz (NE) 4, 7, 10.

28.Leibniz (GP) 7, p. 295 f; 194 f.

29.Leibniz (GP) 2, p. 62.

30.Leibniz (GP) 7, p. 300, 301.

31.Cf. (GP) 7, p. 184, 189.

32.For this kind of intensional inclusion I have found a simple interpretation which makes the traditional law "if a is extensionally included in b then b is intensionally included in a" true. Example: animal is extensionally included in living thing iff living thing is intensionally included in animal. Here b is intensionally included in a ($b \sqsubseteq a$) is defined as: $b \sqsubseteq a \leftrightarrow$-$(\forall z)(b \subseteq z \;\; a \subseteq z)$. Cf. my (CEM), p. 333. This definition can be incorporated into the theory of (virtual) classes and leads to dual theorems (with intensional inclusion).

33.Cf. Leibniz (OF), p. 77-84.

34.For further detail cf. my (IMS), p. 172 ff. and Lukasiewicz (ASy) and Marshall D. (LLA).

35.Leibniz (GP) 1, p. 57.

36.Leibniz (OF), p. 161.

37.Cf. Hermes (ILG), p. 98.

38.Russell (PM), p. 5.

39.Frege (GGA), Introduction p. VII.

40.Russell (IMP), p. 202 f.

41.Russell (IMP), p. 203 f.

42.Gödel (RML), p. 146.

43.This was pointed out by Wang. Cf. Wang (FMP), p. 234.

44.Cf. Kreisel (ACM), chapter 4.

45.There are good reasons to count the axioms of set theory (except that of extensionality) to mathematics according to the following simple criterion of demarcation: In all these axioms the existence of certain sets are asserted - the paradigm-case is the axiom of comprehension - whereas for axioms of logic one would have to require possible universalization of all the variables. Cf. my (DLM).

46.Bernays (MEW), p. 96.

47.Cf. Kreisel (HPG), p. 173.

48.Ibid.

49.Cf. Hilbert (GEZ), p. 489, 494.

50.Cf. Hilbert (GSA) III, p. 180.

51.Cf. Hilbert (GSA) III, p. 157.

52.Skolem (UVC) and (NCZ). Cf. Kreisel (HPG), p. 212.

53.Cf. Kreisel (HPG), p. 214 and Hilbert (GSA) III, p. 165.

54.Also stressed by Kant as the main task of philosophy. Cf. (KRV).

55.Hermes thinks that the description of the "ars inveniendi" in Leibniz suggests that he had in mind a notion which can be represented in modern terms by "recursive enumerability". Cf. ch. 2.44 above.

56.Cf. (PDS), (APS).

57.Cf. Körner (CTh), Cleave (AEB), Schurz (DRS), Weingartner-Schurz (PSS), Schurz-Weingartner (VDR), Weingartner (APS).

58.Cf. Schurz-Weingartner (VDR).

59.For details cf. Weingartner-Schurz (PSS), Schurz (RDS), Weingartner (RCT), (APS) and Schurz-Weingartner (VDR).

REFERENCES

Bernays, P. (MEW) "Mathematische Existenz und Widerspruchsfreiheit", in: Bernays, P.: *Abhandlungen zur Philosophie der Mathematik.* Darmstadt 1976, pp. 92-106.

Cigler J./-Reichel H.-C.: *Topologie. Eine Grundvorlesung.* BI Hochschultaschenbücher, Bd. 121. Mannheim-Wien-Zürich ²1987.

Cleave (AEB): "An Accont of Entailment Based on Classical Semantics", *Analysis* 34, pp. 118-122.

Descartes (RD): *Rules for the Direction of the Mind.* Indianapolis 1961.

Frege, G. (GGA): *Grundgesetze der Arithmetik,* Bd. I Jena 1893 (Darmstadt 1962), Bd. II Jena 1903 (Darmstadt 1962).

Gödel, K. (RML): "Russell's Mathematical Logic", in: Russell, B.: *The Philosophy of Bertrand Russel,* ed. A. Schilpp, New York 1944, pp. 125-153.

Hermes, H. (ILG): "Ideen von Leibniz zur Grundlagenforschung: Die ars inveniendi und die ars iudicandi", in: *Studia Leibniziana Supplementa,* Vol. 15-4, K. Müller, H. Schepers, and W. Totok (eds.), Wiesbaden 1975.

Hilbert (GEZ): "Die Grundlegung der elementaren Zahlenlehre", in: *Mathem. Annalen* 104 (1931), pp. 483-494.

Hilbert (GSA): *Gesammelte Abhandlungen.* Berlin 1970.

Kant, I. (KRV): *Kritik der Reinen Vernunft.* Riga 1781 (1787).

Körner, S. (CTh): *Conceptual Thinking.* New York 1959.

Kreisel, G. (ACM): "On Analogies in Contemporary Mathematics", to appear.

Kreisel, G. (HPG): "Hilbert's Programme", in: Benecerraf, P. and Putnam, H. (eds.): *Philosophy of Mathematics.* Englewood Cliffs 1964, pp. 157-180

Leibniz (GP): *Grundgesetze der Philosophie.* Frankfurt 1962.

Leibniz (NE): *Nouveau Essais sur l'entendement humain.* Frankfurt 1961. Leibniz (OF): *Opuscules et fragmetns inédits de Leibniz.* Paris 1903.

Lukasiewicz, J. (ASy): *Aristotle's Syllogistic from the Standpoint of Modern Formal Logic.* London ²57.

Marshall, D. Jr: (LLA): "Lukasiewicz, Leibniz and the Arithmetization of the Syllogism", in: *Notre Dame Journal of Formal Logic* 18(1977), pp. 235-242.

Popper, K.R. (OUN): *The Open Universe.* Totowa 1982.

Rescher, N. (LIP): *Leibniz. An Introduction to His Philosophy.* Oxford 1979.

Russell, B. (IMP): *Introduction to Mathematical Philosophy.* London 1919 (1960).

Russell, B. (PM):*The Principles of Mathematics.* London 1937 (1903).

Schurz, G. (DRS): "Das deduktive Relevanzkriterium von Stephan Körner und seine wissenschaftstheoretischen Anwendungen", *Grazer Philosophische Studien* 20 (1983), pp. 149-177.

Schurz, G. (RDS): "Relevant Deduction. From Solving Paradoxes Towards a General Theory", *Erkenntnis* 35, 1991.

Schurz, G., Weingartner, P. (VDR): "Verisimilitude Defined by Relevant Consequence Elements. A New Reconstruction of Popper's Original Idea", in: Kuipers, Th. (ed.): *What is Closer-to-the-Truth?* Amsterdam 1987, pp. 47-77.

Skolem, T. (NCZ): "Über die Nicht-Charakterisierbarkeit der Zahlenreihe mittels endlich oder abzählbar unendlich vieler Aussagen mit ausschließlich Zahlenvariablen", in: *Fund. Math.* 23 (1934), pp. 150-161.

Skolem, T. (UVC): "Über die Unmöglichkeit einer Charakterisierung der Zahlenreihe mittels eines endlichen Axiomensystems", in: Skolem, T.: *Selected Works in Logic* (ed. J.E. Fenstad), Oslo 1970, pp. 345-354. (Originally published in 1933)

Wang, H. (FMP): "The Formalization of Mathematics", *Journal of Symbolic Logic* 19 (1954), pp. 241-266.

Weingartner, P. (APS): "Antinomies and Paradoxes and their Solutions", in: *Studies in Soviet Thought* 39 (1990), pp. 313-331.

Weingartner, P. (CEM): "On the Characterization of Entities by Means of Individuals and Properties", *Journal of Symbolic Logic* 39 (1974), pp. 323-336.

Weingartner, P. (DLM): "On the Demarcation between Logic and Mathematics", *The Monist* 65 (1982), pp. 38-51.

Weingartner, P. (IMS): "The Ideal of Mathematization of All Sciences and of 'More Geometrico' in Descartes and Leibniz", in: Shea, W. R. (ed.): *Nature Mathematized.*

Proceedings of the 3rd International Congress of History and Philosophy of Science (Montreal 1980). Dordrecht 1983, pp. 151-195.

Weingartner, P. (PDS): "A Proposal to Define a Special Type of Proposition", in: Weingartner,, P./Schurz, G. (eds.): *Philosophy of the Natural Sciences. Proceedings of the 13th International Wittgenstein-Symposium, Kirchberg/Wechsel*, Austria 1988. Wien 1989, pp. 348-353.

Weingartner, P. (RCT): "Remarks on the Consequence-Class of Theories", in: Scheibe, E. (ed.): *The Role of Experience in Science. Proceedings of the 1986 Conference of the Académie Internationale de Philosophie des Science (Bruxelles) Held at the University of Heidelberg.* Berlin 1988, pp. 161-180.

Weingartner, P., Schurz, G. (PSS): "Paradoxes Solved by Two Simple Relevance Criteria", in: *Logique et Analyse* 3 (1986), pp. 3-40.

consequent idea that the only alternatives to reductionism are the non-scientific vitalistic opinions. In any case a sharp analysis of reductionism in biology must be carried out, also because nothing proves that the alternative quoted above is valid.

However, today in the discussions on reductionism in biology there are many ambiguities. The most significant is that the term «reductionism» in biology is often used in different senses, which are not always clearly defined.

The author of this paper is not a philosopher, or a historian, or a logician, but merely a biophysicist. His opinions have been developed looking around biologists at work, listening to their discussions and ideas. My first concern is therefore to do a sort of census of opinions really shared by biologists working in the laboratories. In so doing, I believe I can distinguish at least three different meanings of the term «reductionism», as used by biologists, as a matter of fact. Of course, in the present paper, references will be made only to biologists really dealing with real biological problems.

The three meanings of the term «reductionism» are as follows. The first, we can define as a «formal reductionism», means that the biological theories could be deduced from the principles of other sciences: physics and chemistry. The second, we can call a «factual reductionism», means that many biologists agree that physical and chemical laws are in principle sufficient to give account of structures and functions of the biological systems. The third one, we can call a «methodological reductionism», means that the methods that have so efficiently worked in physics, should work as efficiently in biology. We must analyze the value of these three meanings of the term «reductionism».

1 - FORMAL REDUCTIONISM.

This meaning of the term is accepted and maintained by few biologists, generally interested in the problems of philosophy of science and logic.

There is, in fact, a very weak tradition of biologists interested in the use of language in biology. This tradition began with the well-known book by Woodger[2].

At present the main utiliser of this very rigorous and formal meaning of the term is Francisco Ayala, the well-known geneticist, working in California. He wrote some time ago an article in AMERICAN SCIENTIST[3] in which he defined «reduction» as «... the explanation of a theory or a set of experimental laws established in an area of inquiry, by a theory usually formulated for some other domain ». Quoting Nagel[4], he stated two formal conditions that must be satisfied to effect the reduction of one science to another one: the *condition of derivability* (all the experimental laws and theories of the «reduced» science must be shown to be logical consequences of the theoretical constructs of the primary science) and the *condition of connectability* (if the laws of the «reduced» science contain some terms that do not occur in the primary science is necessary to establish suitable connections between the terms of the two sciences through a redefinition of the terms of the secondary science, using terms of the primary science). Ayala had no difficulty in showing that genetics, as an example, or cellular theory and so on, cannot be logically deduced by quantum mechanics or other physical theories, nor terms as «gene», «chromosome», «cell», and so on can be translated into terms such as electrical charge, covalent bond, elementary particle and so on. I do not want to discuss here the details of the subject because they can be seen in the quoted article. But I can add that the «reduction» of the basic terms of biology to terms of physics, for example, also in the simpler cases, presents insurmountable obstacles. I tried to make some attempts in this direction, but with no success. The reason probably is that the real object of biology is not atoms and molecules, as complicated as they can be, objects of physics and chemistry, but their ordered configurations which can give rise to complex structures.

Of course one could try to move back toward a sort of soft formal reductionism. As has been noted[5], the way to reduce a theory T1 to a more general theory T2 is not only the inclusion of the sets. In principle two different theories (T1 and T2) can give account of the same facts. In this case from an experimental point of view the two different theories are both true. However, T1 $\wedge$ T2 is not true from a logical point of view. In this sense one could say: biology cannot be reduced to physics in the sense described by Ayala, they are different sciences, with different domains. But they are not contradictory sciences. Biological systems follow physical laws, their atoms and their molecules verify quantum mechanics, the laws of thermodynamics are satisfied, and so on. Biology is not included in physics, but the two sciences partially overlap. Naturally this is true. There is no doubt that a chlorophyll molecule, when studied by the technique of X-ray diffraction, behaves like whatever other molecule. But the point is that the resolution of the structure and of the chemistry of the chlorophyll molecule does not exhaust any biological problem connected with the chlorophyll molecule as a component of a complex biological structure. I mean: a soft formal reductionism can affirm reasonable statements, but it is no more a «reductionism». It is only an assertion of the fact that any experimental activity in the field of science is, by definition, an experimental activity done on the material nature, i.e. on atoms and molecules. This is not «reductionism», it is possibly tied to another problem: that of the unity of the nature.

I think we can conclude that «formal reductionism» cannot be defended, at least for what concerns biology.

2 - Factual Reductionism.

As has been said, many biologists believe that physics and chemistry can give biology a sufficient background to explain the structures and functions of biological systems. Of course biology has reached, on this basis, important goals in recent years. Molecular biology, in the sense of the molecular genetics, DNA activity and coding, genetic engineering, and so on, but also in a broader sense: electrical behaviour of a neuron, interaction between a photon and a molecule of chlorophyll, and so on, is an exciting example of the progress made by new biology. Nevertheless many problems of the biological sciences still need some explaining. We cannot explain why a fertilized egg develops to reach its final form, exactly determined and extremely complicated by sophisticated structures, we don't know how exactly a complex neural net functions to give the elaborated behaviour we can verify. In particular we don't know why and how the circulation and the elaboration of the action potentials in our brains transform themselves into the qualitative perception of colours and sounds. The argument generally used is that the discoveries already made are in favour of the idea that when we do know everything we have to know about a fact, such a fact will be accounted for by the laws of physics and chemistry. In other words, there are no doubts that the processes and the structures considered at a particular moment will be sooner or later «reduced» to some physical principle.

In these tenets there is some truth. On the basis of the knowledge of physics, biology in recent years has made enormous progress. There are no reasons to ignore a method which has given such brilliant results and which can work again and again.

Naturally, it cannot be excluded that in some cases this method can fail. The affirmation that physical and chemical laws are sufficient to account for biological structures and functions, seems then to be a more legitimate affirmation of a program of research, than the statement of the general principle of reductionism.

To be justified in asserting the general principle of reductionism, we would have to be certain that the method described above always functions. How can we reach such a certainty? Or it is only a question of faith? As a matter of fact there is a possibility that faith becomes not a certainty, but a fact of high probability. This possibility, however, is tied to a particular assumption on the nature of biological systems. Such systems are made by atoms and molecules and the behaviour of atoms and molecules is described very well by physical laws. To simplify: if in biological systems there is nothing else than atoms and molecules, the reduction of biology to physics can be considered highly likely. In other words: this kind of reductionism demands a combinatory interpretation of the nature of biological organisms. This opinion seemed for some time a reasonable one. Let us consider genetics. Genes determine the phenotypical characters of organisms, and genes combine at random, giving origin to a particular organism. In the classical evolutionary theory mutations are random events and they give origin to different organisms. Natural selection then will select the types well fitted to the different environments. But also in terms of molecular biology combinatory views seemed to work. In the primitive seas the amino acids combined randomly, giving rise to different proteins, some of which work. Moreover, the spatial structure of protein molecules determines their spatial arrangements and hence the birth of the little cellular organs. The argument can be generalized to a general combinatory theory of biological organisms.

But now this view has shown many difficulties. Genetical structure appeared much more complicated than we had believed. And we know that any combination of genes cannot be accepted. There are constraints of internal coherence which are difficult to define, and which, however, cannot be eliminated. In addition I presented recently at a meeting[6] some calculations to show that, in the framework of the combinatory view, the possible combinations are so numerous (of the order of 10^{100}), that the probability of getting the desired structure at a given time is practically negligible (given the fact that the age of the earth is only of 10^{17} sec). Everything seems therefore to be in favour of the idea that we have to consider also some principle of order that can act at the higher levels of the organization of matter. Of course we are not thinking of some old vitalistic principle as we will try to clarify in the conclusions of this paper.

With the combinatory hypothesis being probably not correct, also the reductionist thesis we are here discussing seems dubious. In addition this reductionist thesis seems to meet also some logical difficulties. In fact, the thesis, as we have seen above, affirms that when we do know «everything» about a fact, such a fact will be certainly «reduced» to some physical principle. But, what does it mean to know «everything» about a fact? There is a sort of a vicious circle, because, analyzing carefully the argument, one discovers that, in this context, we are authorized to say that we know «everything» about a fact when we can «reduce» the fact to some physical principle.

This point would demand detailed discussion. To shorten the analysis we can schematise the argument in terms of system theory. Consider a system or a sub-system. It can be an organism, or an organ of the organism, or, in general, a structure able to perform some process. It will receive from the external environment some signal the effect of which will be that the system will perform some process, regulated by the structure of the system itself. The process will act in some way on the environment. The situation could be described in general by the schema in the figure.

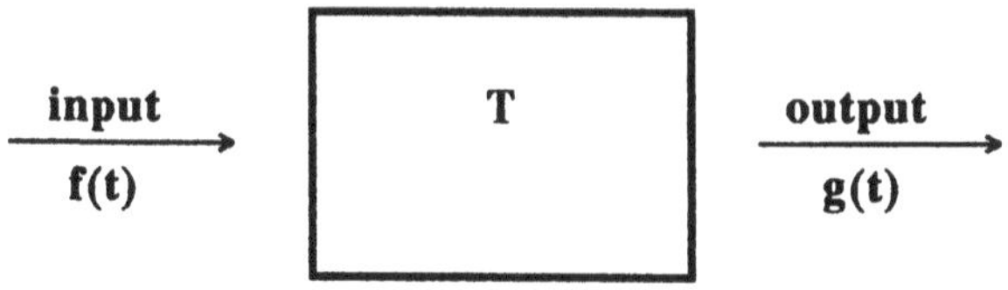

The signals entering the system are the inputs and they could be described, in principle, by a function of time f(t). The system operates following its own rules and the consequence will be the outputs, which could be described too, in principle, by another function of time g(t). If we know «everything» about the system (the box illustrated), we know the rules governing the operations of the box. In this case, knowing f(t), we can deduce g(t). Mathematicians know that to understand the rules governing the box, means to know the «transfer function T» [the function that applied to f(t) gives g(t)]. We can say that to know «everything» about the system means to know the transfer function T. If it is so we can deduce the outputs from the inputs. We have «reduced» the working of the system to their rules (physical and chemical rules governing the behaviour of the molecules constituting the system). The point is: we say that «reduction» (to deduce g(t) from f(t)) means to know T. But we can say that we know T when the «reduction» is made. But when we cannot deduce g(t) from f(t), what can we say? That «reduction» is not possible, or that we still do not know T? The answer seems to be determined only by the philosophical opinion of the observer.

Of course when it is possible to deduce g(t) from f(t) there are no problems. In biology in many cases this situation occurs. But in other cases it is not so. In these more complicated situations we can say nothing. The possibility of a «reduction» remains indefinite. If one, for general reasons, believes in the truth of reductionism, he will probably say: the day will come in which we will know so many things that reduction will be demonstrated. In the case of very complex systems also this assertion, rooted in faith, can be proved false. Take as an example a human brain. This is our system: a sort of black box. Much information enters the box in time. The brain works, elaborates the information and gives out some information, determining some behaviour on the external world. Now «reductionism» means that if I could know in detail the electrical behaviour of the neural nets, interpreted eventually as logical net, and if I know the inputs, I could deduce the outputs.

The problem is formally the same as that presented above. When, knowing a large number of rules governing the working of the brain, I will not be able to deduce the outputs from the inputs, what will I say? That «reductionism» is not valid and that there are some residues I cannot explain: a sort of sixth proof of St. Thomas? Or will I say that I don't know at present all the rules guiding the brain? I would be forced to give the second answer, given the fact that I have no other criteria to say that I know all the implied rules, excepting the possibility of deduction of the outputs from the inputs. So I can hope that in the future, sooner or later, I will be able to deduce the outputs from the inputs, knowing «all» the rules of the brain. But, what is the probability that matters are really going to take this course? With reference to the schema in the figure, what are the conditions necessary to be able to deduce g(t) from f(t)? Of course the first condition is to know the transfer function T, i.e. the rules of the box. But there is another condition: we have to deal with a deterministic system. Only in this case, in fact, are the outputs univocally determined by the inputs and by the rules guiding the working of the box. If the system were not deterministic the situation could change drastically. The presence in the system of some probabilistic gaps could create particular conditions in which the outputs are no longer the simple univocal effect of f(t) and of the rules of the box. In these cases it is possible that, knowing the inputs and «all» the rules of the box, and excluding any additional residue, nevertheless the outputs are not deducible with certainty. Now the criterion of ascertaining our complete knowledge of the rules of the box as a function of the possible deduction of the outputs vanishes. And consequently the assertion of reductionism loses all sense, since we cannot say in any cases if the non-deducibility of the outputs is caused by our incomplete knowledge of the rules of the box, or by the non- deterministic nature of the system, or by other causes.

As concerns the example of the human brain, we don't know, of course, «all» the rules of the box, but we know that such a piece of machinery is certainly a non-deterministic one. The conclusion is that also «factual reductionism» cannot be maintained, at least as a general view of biology. It becomes, in fact, only a belief. It can survive, of course, as we have said, only as a limited program of inquiry, restricted in particular to biological molecules or molecular structures which can be studied only in terms of physical and chemical laws.

3 - METHODOLOGICAL REDUCTIONISM.

The situation described in the two preceding paragraphs drove many biologists to share a sort of soft reductionism: a methodological reductionism. The methods of research in biology have to be «reduced» to those so efficiently used by physics. Experimental methods, the use of mathematical models and deduction from them of consequences experimentally testable, and so on, have to be utilized by biologists too. I agree on the last point and I think that everybody could also agree on that. But I have some doubt that the quoted sentence is really an affirmation of reductionism. In fact, we are speaking of «reduction» to what? To the method of the natural science, naturally. But, to say that we «reduce» the method of one particular science to the method of science «tout court» is, in the best case, a tautology.

The assertion that the method of biology must be the method of science, although strengthened by the considerations presented at the end of the preceding paragraph, poses some problems. It implicitly affirms that biology is a science and this assertion seems obvious. But, what kind of science is biology? Some philosophers think that biology is a historical science. Some physicists, such as Ageno[7], tend to share this conviction. History is the science of the contingent. Historians ascertain that particular events have occurred, but they know that such events were not necessary: they cannot built up a general theory to foresee such events. When we think, as an example, of biological evolution, it seems to be in the same situation. The fact that the horse appeared in a given geological era seems a contingent event. But the «hard» sciences deal with some kind of necessity. See, as an example, the laws of motion in physics. So it is possible elaborate theories. How is it possible to have a real theory of evolution?

In other words the true problem of contemporary biology is if true and autonomous biological laws are possible, not if biological laws are reducible to physical laws. This problem, however, has nothing to do with the problem of the reductionism.

4 - Conclusions.

The analysis just performed seems to lead to the conclusion that, in the present state of biological research, the concept of reductionism is not an operational instrument and it could create many misunderstandings, drawing the attention of scientists far from the real problems of biology. Reductionism, in fact, is not a tenable thesis, like formal reductionism, a sort of faith, as factual reductionism is, or a simple tautology as in methodological reductionism.

As we said above, a central point in the discussion of the general problems of contemporary biology is the question if biology can establish mathematical models and scientifical theories, and if so, what kind of theories.

We cannot discuss here such a problem because it is a very difficult one and it only skims the problem of reductionism, we are discussing in this paper. For what concerns the arguments presented here, it is sufficient to add that probably we have to deal with the problems of complexity, in the sense that we have at present a very limited knowledge of the constraints acting on dynamic structures, and of the rules regulating the complex interactions between «parts», to build up the «whole» organism[8]. In this framework, the fact that today the physical sciences too are encountering concepts such as complexity, structure, ordered systems, and so on also has to be remembered. I am referring to new thermodynamics (particularly non-linear irreversible thermodynamics), to the theories of dynamic systems, to fractal geometries, to the theory of catastrophes, and so on. In this context «reductionism» is no longer necessary. Paradoxically the ideas of Simpson[9] that physics could be «reduced» to biology, acquire a strange significance, as a sign of some «laws of complexity».

To conclude: expressions such as «...organisms are alleged to be «wholes» that must be studied as wholes and not as «sum» of isolable parts...»[10], considered old and, in a certain sense, meaningless, could be recovered, although in a new and more accurate meaning.

Department of Physics - University of Genoa (Italy)

REFERENCES

1) J.O. de Lamettrie: *L'Homme Machine*. Leida, 1748.

2) J.H. Woodger: *Biology and Language*. Cambridge University Press, Cambridge, 1952.

3) F.J. Ayala: «Biology as an autonomous science». In: *American Scientist*, *56*, 3, 1968. Pag. 207.

4) E. Nagel: *The Structure of Science*. Harcourt, Brace and World, New York, 1961.

5) M.L. Dalla Chiara , G. Toraldo di Francia: *La scimmia allo specchio*. Laterza, Bari, 1988. Pag. 123 and following.

6) R. Morchio: *What is the meaning of biological evolution in Cosmology?*. Proceedings of the Meeting: «La Cosmologia nella Cultura del '900». In preparation.

7) M.Ageno: *Le radici della biologia*. Feltrinelli, Milano, 1986.

8) R. Morchio: «Biophysics and Research on Structures». Abstracts of «I° Convegno sulle Strutture Dinamiche». *Rivista di Biologia - Biology Forum*. In press.

9) G.G. Simpson: *This View of Life*. Harcourt, Brace and World, New York, 1964.

10) E. Mayr: «Cause and Effect in Biology». In: D. Lerner (ed.) *Cause and Effect*, Free Press, New York, 1965. Pag. 33.

Hans Primas

REDUCTIONISM: PALAVER WITHOUT PRECEDENT

Introduction

Reductionism, if accepted, is usually accepted on faith and without logical evidence or sound reasons. For example, overblown claims in the philosophical literature for the reducibility of thermodynamics or chemistry to Hamiltonian mechanics or quantum mechanics are not justified by present scientific knowledge[1-5]. Indeed, I do not know of a single scientifically interesting example of a relation between theories which fits any of the schemes for theory reductions proposed by the philosophers of science, and which fulfills the legitimate requirements of rigor of scientists.

Very vaguely, reductionism claims that all the laws of higher levels can be explained by those of the lowest und most fundamental level so that nothing intrinsically new enters at the higher levels. For example, it has been maintained that "all theories of biology can in principle be reduced to physical theories". Of course, if reductionism is understood merely as a *heuristic tool* or even as a *research strategy*, there are no objections. Wild speculations are of utmost importance for every creative scientist, so the question can only be whether reductionistic speculations are wild enough. Similarly, research strategies may be sucessful or bad, but all such questions have nothing to do with the philosophical problems related to

E. Agazzi (ed.), The Problem of Reductionism in Science, 161–172.

reductionism. An appropriate explication of the concept of "reduction" should transform the original heuristic idea into a *nontrivial and scientifically vindicable* form. In order to avoid muddled arguments, I restrict my discussion to *theory reductions*, and consider only well-defined theories that are logically consistent and empirically reliable. Furthermore we have to distinguish at least three variants:

Strong reduction:

The higher-level theory together with its interpretation can be deduced from the basic theory. No non-trivial interdisciplinary example is known.

Weak reduction:

The higher-level theory can be deduced from the basic theory together with appropriate initial and boundary conditions.

Compatibility:

The laws of the fundamental theory (e.g. quantum mechanics) *apply fully to all higher level descriptions* (like chemical or biological phenomena). If no additional claim is added –neither in the sense that these laws already suffice to explain higher-level phenomena, nor in the sense that additional (say vitalistic) laws exist - this mere compatibility requirement is nowadays accepted by most scientists.

Strong reduction is an uninterestingly rigid requirement, while weak reduction is to a considerable extent subject to ad hoc interpretations and leaves too many loopholes open to be of scientific interest. Compatibility can be considered as a trivial minimum requirement since present-day physics is flexible enough to include almost any conceivable new fundamental law. According to my view, there remain two paramount questions:

(i) How do we have to define sensibly and precisely what is meant by theory reduction?

(ii) Are there processes which obey the fundamental laws of physics, but nevertheless cannot - even in nciple - be explained by the known fundamental laws (or by suitable universally valid generalizations) alone?

Critics on alleged theory reductions

In every serious discussion of theory reduction we have to insist that the higher-level theory must be deduced *rigorously* and *completely* from the basic theory, indicating within the adopted mathematical formalism *precisely all necessary auxiliary conditions.* Since the philosophical literature on reductionism is teeming with scientific nonsense, I insist that every assertion of a successful reduction requires an *explicit proof.* The following examples show that one cannot discuss the problem of theory reduction in science if one knows science only by hearsay or from the popular literature.

First illustration: Chemistry has not been reduced to physics

In their famous paper on reductionism, Kemeny and Oppenheim[6] write: "*...a great part of classical chemistry has been reduced to atomic physics*". No references for a proof are given. Evidently, the authors have no idea what they are talking about. *Queries*: How are the non-linear differential equations of chemical kinetics derived from linear quantum mechanics? Does there exist a universal relation between chemical substances and molecules?

Second illustration: Chemical purity is not a molecular concept

Mario Bunge[7] claims that the proper definition of chemical purity can be phrased as follows: "*Substance x is chemically pure if and only if x is composed exclusively of either atoms or molecules of a single species or kind*". This is a classical category mistake since it tries to explain the purity concept used for chemical *substances* by the purity concept relevant for a category of a different logical type. *Query*: Is liquid water a chemically pure substance? If so, which molecules liquid water is consisting of ? Most certainly liquid water is not composed out of molecules H_2O. A correct discussion can be found in the lectures on thermostatics by van der Waals[8] (compare §5, §92, §93).

Third illustration: The theory of heat has not been reduced to statistical mechanics

Kemeny and Oppenheim[6] claim: "... *the classical theory of heat has been reduced to statistical mechanics*". *Queries*: How can the zeroth law of thermodynamics be derived from statistical mechanics? Why are there no Einstein-Podolsky-Rosen correlations in thermodynamics?

Fourth illustration: Temperature is not a molecular concept

Ernest Nagel[9] claims that the concept temperature can be reduced to statistical mechanics by a *mere definition*, namely via the relation $E = \frac{3}{2} k\mathrm{T}$ as the mean kinetic energy E of gas molecules. This example - cited again and again - bespeaks an incredible ignorance of most elementary concepts of physics.

Fifth illustration: Classical physics is not a limiting case of Einsteinian physics

One often reads that "pre-Einsteinian physics is the limiting case of special relativity theory for infenitely large velocity of light".This cannot be so since classical physics is a consistent theory which includes Newtonian mechanics as well as Maxwell's electrodynamics. However, since no unique Galilean limit exists for Lorentz-covariant electromagnetic interactions[10], pre-Einsteinian physics cannot be reduced in a simple way to Einsteinian physics.

Sixth illustration: Classical physics is not a universal limiting case of quantum physics

There is a widely held belief that classical mechanics is the limiting case for vanishing Planck's constant *h*. However, it is easy to give counterexamples, and to prove rigorously that *there is no universal classical limit of quantum mechanics* (i.e. the limit $h \to 0$ does not exist in the norm topology). What is possible is that for distinguished small families of quantum states certain quantum systems may behave exactly as classical systems. In this case we speak of classical *quantum systems*[11-14], which, of course, depend on the physical value of Planck's constant *h*. However, whether a quantum system behaves classically or not, is *not* an intrinsic but a *contextual* property of the system. Depending on the context, it is possible to associate to one and the same quantum system various *inequivalent* classical quantum systems.

Reduction to first principles of physics?

According to our present understanding of the first principles of quantum mechanics, *atomism as well as reductionistic-mechanistic philosophies have no longer any scientific basis*. The idea that in the final analysis, there should exist only one fundamental description of nature is in striking contradiction to the ideas of quantum theory. Modern quantum mechanics compellingly demands a *multitude* of context-dependent descriptions. Atomism is dead, the material world certainly is not constructed out of independently existing elementary systems . Only under quite special conditions, a description by context-dependent entitites such as quarks, electrons, atoms or molecules is allowed. Accordingly, the ideas of reductionism and atomism should not become muddled together. Questions like "Do organisms exhibit properties other than those of their constituent atoms and molecules?" [15] are ill-posed since most certainly organisms are *not made out of atoms and molecules*, even if it should turn out that all properties of organisms can be explained in terms of quantum mechanics. Already the lack of individuality of electrons disqualifies them as building blocks in the sense of Newton's idea that "God in the Beginning form'd Matter in solid, massy, hard, impenetrable moveable Particles _" [16]. That is, in the discourse on reductionism we have to avoid naive atomistic ideas. The assertions

> "something is made out of elementary systems",
> "something can be decomposed into elementary systems",
> "something has an explanation in terms of elementary system",

are not equivalent. According to quantum mechanics the first one is almost always false, in spite of the fact that the second one is often true, and that the third one may be sensible.

Intertheoretical relations

Kemeny and Oppenheim[6] think that "it is hopeless to attempt a rigorous logical analysis of science in the form in which we normally find it. It is not the role of the working scientist to make his fundamental assumptions clear, and it is not reasonable to expect that he will proceed according to rigorous logical rules". This statement only shows that the authors do not know the literature. Besides, the niveau of the scientific discourse adopted by these philosophers is absolutely unacceptable. The standard we have to require for any serious study of theory reduction has been set by Friedrichs[17] in his analysis of the relation between geometrical optics and electrodynamics. Scientists have studied many intertheoretical relations in a mathematically rigorous way and found conclusively that *none of the well-investigated intertheoretical relations correspond to any of the reduction relations defended by the philosophers*. Most intertheoretical relations are mathematically describable as *singular* asymptotic expansions - a mathematically very delicate and demanding undertaking. Asymptotic expansions are never universally convergent but valid only in a very specific context. In such context-dependent descriptions the specification of the context is at least of equal importance as the first principles. *Given the context*, the new description can be derived from the basic principles but the context is not given by the first principles of physics. In other words, these new higher-level descriptions are *compatible* with the first principles, but cannot be derived from them alone.

The context-dependence of any description of nature is intrinsically related with the fact that we can say nothing about nature without making abstractions. We have to abstract from irrelevant features, but it is not laid down by natural laws what has to be considered as relevant and what as irrelevant. *Without abstractions, there is no science.* The abstractions we introduce into experimental sciences have also to be introduced into a theoretical description by corresponding new *contextual topologies* which as a rule induce *symmetry breakings*. Different branches of science are using different approaches which can be characterized by different topologies and symmetry breakings.

Emergence of novelty by symmetry breaking

A mathematically rigorous deduction of emergent quantities from a basic theory requires a breaking of fundamental symmetries, either by a singular asymptotic expansion which does not converge in the topology of the primary theory, or by choosing one of the infinitely many possible, physically inequivalent representations of the primary theory (Gel'fand – Naĭmark–Segal-construction of algebraic quantum mechanics). In both cases, a *new topology* has to be introduced by our preferences, which are either determined by our own cognitive apparatus, or by the abstractions of the "pattern recognition devices" connected with the experimental equipment.

As a rule, symmetry breakings induce the emergence *of hierarchical levels* with clearly separated time ranges. But it would not be correct to say that nature is structured hierarchically. It is our *viewpoint with the associated abstractions* which generate the higher levels of a hierarchical description. *The task of higher-level description is not to approximate the fundamental theory but to represent new viewpoints.* However, *the universally valid first principles are never sufficient for deducing higher-level theories.* Nevertheless, *given a well-specified context*, higher-level descriptions often can be derived by asymptotic expansions. These expansions are not uniformly convergent but one can achieve convergence with respect to a *new* topology in a restricted domain. The closure of the higher-level description in this new topology introduces automatically new context-dependent observables which are not already present in the fundamental description. In this mathematically precise sense, one can speak of the *emergence of novelty* in descriptions on a hierarchically higher level.

Examples for emergent quantities in physical descriptions

First example: Shadows

The phenome passes an object does not exist in Maxwell's electrodynamics since this fundamental theory is governed by partial differential equations which have *continuous* solutions. The discontinuities associated with shadows emerge via Stokes' phenomenon in a singular asymptotic expansion of Maxwell's equations leading to geometrical optics[17]. It would be misleading to consider geometrical optics as an approximation to Maxwell's theory. They are related but *qualitatively different* theories.

Second example: Molecular structure

The concept of molecular structure as used by chemists is *compatible* with the first principles of quantum mechanics, but does not follow from them (compare the review given by Weininger[18]). Historically, it has been introduced into quantum chemistry by the so-called Born-Oppenheimer approximation. However, this terminology is very unfortunate because the main issue is not an approximation but the breaking of the holistic symmetry of quantum mechanics. The Born-Oppenheimer-description is a typical example for an asymptotic weak reduction.

Molecular structure is an emergent quantity, quantum-mechanically described by *new classical observables* which are generated by the abstraction of the actually existing EPR-correlations between electrons and nuclei[2,19].

Third example: Quantum mechanics vs. classical mechanics

According to the modern point of view, quantum mechanics is a fundamental theory while classical mechanics is a higher level theory of limited validity. According to this view, all classical systems are in fact *classical quantum systems*, characterized by a commutative algebra of observables and a Boolean logical structure. All classical quantum systems depend on Planck's constant h, they emerge by breaking the holistic Einstein-Podolsky-Rosen symmetry, not by the limit $h \to 0$. Classical observables are elements of the center of the W*-closure of an appropriate Gel'fand-Naĭmark-Segal-representation of the C*-algebra of intrinsic observables. Classical observables are truly *emergent* quantities: they can be constructed from first principles only by *selecting* one of the infinitely many inequivalent representations of the fundamental C*-algebra. The selection is, however, not given by first principles.

Conclusions

There exists not a single scientifically well-founded and nontrivial interdisciplinary example for a theory reduction in the sense of Hempel, Oppenheim, Nagel, Sneed or Stegmüller. Therefore, for a contemporary theoretician reductionism is dead. Clearly, intertheoretical relations are of central interest to modern science but they scarcely have something in common with the old ideas of atomism and reductionism. The traditional concept of theory reduction rests on much too simple a view about the structure of scientific theories. The relationship between different levels of description is much more complicated than the simpleminded ideas of the advocates of reductionism or antireductionism presuppose. Though the reductionist thesis is unwarranted, this fact does not imply an antireductionist thesis.

Traditional theory reduction tacitly presupposes that subtheories can always be totally ordered. This may be true for classical theories but is wrong for quantum theories. It can be shown that the *hierarchy of theories* has the structure of a *non-Boolean* lattice, so that scientific theories can only be partially ordered[20]. Furthermore, all subtheories of a universally valid basic theory are in a state of *peaceful co-existence* with the postulated first principles, *but they cannot be deduced from them without introducing crucial new structural elements*. In contradistinction to fundamental theories, all higher-level theories are context-dependent. Higher-level theories do possess a certain autonomy and cannot be deduced from universally valid first principles without taking into consideration the pattern recognition devices necessary for the observation of high-level phenomena. Any mathematically formulated reduction of a higher-level description to a fundamental theory is conceivable only if in the basic theory a new contextual topology is introduced. This new topology is never given a priori but depends in a crucial way on the abstractions made by the cognitive apparatus or the pattern recognition devices used by the experimentalist. This program can be realized in terms of modern algebraic quantum mechanics. In this framework it is possible to describe the behavior of matter in many, mathematically precisely characterized, mutually exclusive complementary ways[2,21]Each hierarchical level requires an autonomous, non-reducible language which should not be eliminated in favor of an empty "universal language". Mutually exclusive complementary descriptions of nature are not only admissible, but they are equally entitled and necessary. That is, *science is necessarily pluralistic*.

Laboratory of Physical Chemistry, ETH-Zentrum, Zürich (Switzerland)

REFERENCES

1. H. Primas and W. Gans: *Quantenmechanik, Biologie und Theoriereduktion.*
In: *Materie - Leben - Geist. Zum Problem der Reduktion der Wissenschaften.*
Ed. by B. Kanitscheider. Berlin. Dunker & Humblot. 1979. Pp. 15-42.
2. H. Primas: *Chemistry, Quantum Mechanics and Reductionism.*
Berlin. Springer. 1981, second edition 1983.
3. H. Primas: *Kann Chemie auf Physik reduziert werden?*
Chemie in unserer Zeit 19, 109-119, 160-166 (1985).
4. H. Primas: *Can we reduce chemistry to physics ?*
In: *Centripetal Forces in the Sciences* Vol. II.
Ed. by G. Radnitzky. New York. Paragon House. 1988. Pp. 110-133, 139-143.
5. H. Primas: *Biologie ist mehr als Molekularbiologie.*
In: *Die Frage nach dem Leben.*
Ed. by E. P. Fischer and K. Mainzer. München. Piper. 1990. Pp. 63-91.
6. J. G. Kemeny and P. Oppenheim: *On reduction.*
Philosophical Studies 7, 6-19 (1956).
7. M. Bunge: *Treatise on Basic Philosophy.* Volume 7.
Dordrecht. Reidel. 1985, p. 225.
8. J. D. v. d. Waals: *Lehrbuch der Thermostatik. Erster Teil. Allgemeine Thermostatik.*
Bearbeitet von P. Kohnstamm. Leipzig. Ambrosius Barth. 1927.
9. E. Nagel: *The Structure of Science.*
New York. Harcourt, Brace. 1961.
10. M. Le Bellac and J. M. Lévy-Leblond: *Galilean electromagnetism.*
Nuovo Cimento 14B, 217-233 (1973).
11. H. Primas: *Induced nonlinear time evolution of open quantum objects.*
In: *Sixty-Two Years of Uncertainty: Historical, Philosophical and Physical Inquiries into the Foundations of Quantum Mechanics.*
Ed. by A. I. Miller. New York. Plenum. 1990. Pp. 259-280.
12. H. Primas: *The measurement process in the individual interpretation of quantum mechanics.*
In: *Quantum Theory Without Reduction.*
Ed. by M. Cini and J. M. Lévy-Leblond. Bristol. Adam Hilger. 1990. Pp. 49-68.
13. H. Primas: *Realistic interpretation of the quantum theory for individual objects.*
La Nuova Critica 13-14, 41-72 (1990).
14. H. Primas: *Necessary and sufficient conditions for an individual description of the measurement process.*

In: *Symposium on the Foundations of Modern Physics*, 1990.
Ed. by P. Lahti and P. Mittelstaedt. Singapore. World Scientific. 1991.
15. F. J. Ayala: *Biology and physics. Reflections on reductionism.*
In: *Old and New Questions in Physics, Cosmology, Philosophy, and Theoretical Biology. Essays in Honor of Wolfgang Yourgrau.*
Ed. by A. van der Merwe . New York. Plenum Press. 1953. Pp. 525-534.
16. I. Newton: *Opticks*. London. 1704. Fourth edition, 1730.
Reprinted by Dover, New York, 1952, p. 400.
17. K. O. Friedrichs: *Asymptotic phenomena in mathematical physics.*
Bull. Amer. Math. Soc. 61, 485-504 (1955).
18. S. Weininger: *The molecular structure conundrum: Can classical chemistry be reduced to quantum chemistry?*
J. Chemical Education 61, 939-944 (1984).
19. H. Primas: *Foundations of theoretical chemistry.*
In: *Quantum Dynamics of Molecules. The New Experimental Challenge to Theorists.*
Ed. by R. G. Woolley. New York. Plenum Press. 1980. Pp. 39-113.
20. H. Primas: *Theory reduction and non-Boolean theories.*
J. Math. Biology 4, 281-301 (1977).
21. H. Primas: *Chemistry and complementarity.*
Chimia 36, 293-300 (1982).

Gilles-Gaston Granger

MUST A SCIENCE OF ARTIFICIAL INTELLIGENCE BE NECESSARILY REDUCTIONIST ?

1. WHAT ARE WE TO UNDERSTAND BY A "SCIENCE" OF ARTIFICIAL INTELLIGENCE ?

1. Artificial intelligence, a very interesting phenomenon from an historical and epistemological point of view, still remains an ambiguous as a discipline. It was born and developed in the wake of a multitude of widely varying attempts to apply formal reasoning to human behavior. Let us mention some of the major stages to underscore this diversity of origin:

-An essay on logical calculus and nervous activity, by Mc Culloch and Pitts (1943)

-The theory of games of strategy, formulated by Von Neuman and Morgenstern (1944), in its application to Economics

-The idea of cybernetics (Norbert Wiener, 1948)

-The theory of information (Shannon and Weaver, 1949)

-Hebb's propositions concerning "the organization of behavior" (1949)

E. Agazzi (ed.), The Problem of Reductionism in Science, 173–182.

In 1946, a new discipline, whose project was just beginning to emerge, was christened "artificial intelligence" by McCarthy at the Darmouth Congress.

2. Beyond the vast variety of the objects and methods of the various enterprises that preceded and prepared its coming, it is nevertheless possible to discern two fundamental traits of the context in which this disciple was born. First there was a *technical background* without which no serious development could have arisen : the formation and rapid progress of a computer technology. In the second place, there was the implicit or explicit influence of the grand *theoretical blossoming* of metamathematics such as that found in the works of Gödel, Turing, and Post. The relation of this second trait to the first has become evident with hindsight, though the second does constitute an independent direction of research that was able to furnish the new discipline with suggestions that were proper to it.

3. At present, it seems as though artificial intelligence was liberating itself from what gives the impression of a syncretic magma of fusion -and confusion-, that of the "cognitive sciences". Under this latter-day aegis are to be found the more or less fruitful and controlled exchanges among robotics, the neurosciences, linguistics and epistemology, though without its being possible yet to discern the constitution of a clearly defined object or project. Should we want to distinguish in all this, insofar as we can, a notion of artificial intelligence in the strict sense, we would say that it takes as its object the phenomena of *representation of the environment*, of *storage and treatment of knowledge*. From this point of view, the word "intelligence" will be taken as referring to the idea of adaptation to an environment and of problem solving : the word "artificial" will refer then to the hypothesis of a reproduction or a simulation of human deeds of intelligence by other means. There can be seen here a triple orientation which is quite ambiguous. First of all there is the properly *scientific* one in that one of avowed tasks *of artificial intelligence* is to construct *abstract models* of certain human

phenomena. In the second place it is technical, since it produces models that should, in principle at least, be materially realizable as automata. And in the final analysis it is *philosophical*, insofar as it addresses the problem of the conditions of the possibility of human acts of cognizance. This last orientation moreover, seems to me to be open to equivocation. In an article that gives a good general presentation of the problem[1] Joelle Proust, following D. Dennett, tries to assimilate this preoccupation with the Kantian concern. The problem here is that artificial intelligence is not looking for what might be called *transcendental* conditions. It seems to me that it is interested only in discovering the *formal* conditions to be imposed on the empirical in order to render the effects of "intelligence" possible[2]. No matter what its interpretation is, the philosophical orientation, whether asserted or only confused, remains constantly present to artificial intelligence along with the scientific and technical orientations that are its driving force. This triple aspect assuredly makes it an ambiguous discipline though -so much so that the question of *reductionism* turns out to be particularly meaningful here. Before addressing it directly though, we should first specify somewhat the very sense of the general question.

2. THE PROBLEM OF REDUCTIONISM IN GENERAL.

We shall propose distinguishing two senses of reductionism:

1. There is first an "ontological" sense which consists in hypothesizing that a type of being or concrete phenomena may be reconstituted in their substrate and in their functioning by means of elements and functions belonging to types of entities or phenomena considered to be more fundamental and more *simple*.

Such hypotheses call for a confirmation on two distinct levels : the empirical and technical on the one hand, the philosophical on the other. In the first case, this will amount to carrying out the reduction by composing the higher level entities and effects simply from elements taken from the lower level. From the second point of view, there is a requisite for criteria that will allow distinguishing, both practically and conceptually, between the orders and the types of entities' in question.

2. The second sense of reductionism is an "epistemological" one. The question that arises here is that of determining whether the abstract models used by one science aren't of the same kind as those used successfully by another, by reference to phenomena considered to be ontologically simpler for example (though this last characterization is not determinant here). It is true that from this purely epistemological point of view one might say that *all scientific knowledge is reductionist*, since every science represents its objects by means of *abstract logico-mathematical models*, thereby, in a trivial sense, reducing them to this fundamental level of knowledge. Clearly what we have in mind however is a stronger sense of reductionism. Even the epistemological reduction would not consist in a simple formal transformation through bringing to light a rather general common structure. It will certainly not be said, for instance, that the community of structure, as witnessed by a second order differential equation with constant coefficients, between the phenomena of the oscillation of current in an electrical circuit and that of the spatial oscillations of weighted masses properly amounts to a reduction of the electrical to the dynamical. A correspondence still more profound must be shown, at another level, between the *contents* of the terms of the two species of phenomena.

The acknowledgement of a genuine reduction depends on a precise sense of the difference between types of models and a hierarchy among these types. If we manage to characterize the irreducibility of different models used by various sciences, we shall have advanced a certain support for a non-reductionist thesis, even though the application of a model to a certain kind of phenomena and entities can be effected only over a material already organized in terms of a still simpler model, but one to which the model in question can nevertheless not be reduced. Granted, such an ordering and intrication among models, perhaps even reciprocal in some cases, would give a weaker sense than the ontological one to the question of reductionism, but it would also be a more positive sense.

It is only in this perspective that the philosopher of science can hope to raise the problem of reductionism in artificial intelligence without bringing to bear too many prejudices. We must therefore examine which models this discipline employs.

3. THE MODELS OF ARTIFICIAL INTELLIGENCE.

We shall not consider the properly technical aspect of the models and their effective implantation in automata, although from a certain point of view this is an issue of capital importance. Yet the success or lack of success of these productions does not depend solely on an *a priori* recognizable philosophical determination of the differentiation of models, but on the observation of empirical results, and more particularly on empirical results yet to come. We shall therefore envision only in a quite general manner the classes of solutions proposed to the problem of the simulation of effects and of the reproduction of functions and aptitudes by means of models in artificial intelligence.

1. First of all, what are the types of phenomena presently dealt with by artificial intelligence? It seems that they can be arranged into four main categories.

a) The representation by an agent (by a subject ?) of its physical environment. Some of the tasks set here for example are those of modeling the recognition of forms, of modeling a behavior of displacement with avoidance of obstacles, or of constructing a model of the neurophysiology of sight and touch insofar as these are productive of images of the milieu.

b) The representation of the "pragmatic" environment of the agent. Here the task may be that of modeling a certain grasping of a complex "scene" that is endowed with sense, that is to say, referring back to implicit and remembered elements. Or again, at a still higher level of complexity, it may be to model behaviors of natural language communication.

c) The storage of knowledge. The attempt here is to model the stabilization of representations, and the processes of reactivation and recall.

d) Lastly, inferences and learning. The models here are to achieve operations by extracting new knowledge from given knowledge, in recognizing and exploiting their connection as in "expert" systems, for example: or again, to improve a certain behavior on the basis of former repetitions.

2. To deal with such phenomena, artificial intelligence seems, at present, to make use of relatively distinct types of models.

In the first place, it is still common to meet with models that are a pure and simple tranposition of actual psychological experience. It is in this vein that J. McCarthy could write, in an article written many years ago, that such models were "legitimate when they express the same information concerning the machine as they express concerning the person" and that in addition they were "useful when they aid in understanding the structure of the machine and the way to repair or to improve it"[3]. Is this not a case of inverse reductionism where the "behavior" of the machine is "explained" by the attribution of fictive "mental qualities" taken from the introspection of human traits ? The same orientation would be found again in the notion of "micro worlds" of "scenarios", and, in a more elaborate fashion, in that of "frames" implanted in the models of recognition of the environment. These different attempts do indeed meet with a certain technical success in the imitation of "intelligent" behavior by a machine. Yet one might ask whether, in following this path, artificial intelligence is not resigning itself to an *epistemological failure*. The pure and simple transposition of mental procedures grasped by introspection must surely meet its limits rather quickly, just as it leaves the problem of the scientific knowledge of these processes completely unresolved : the problem of a psychology of *natural intelligence*.

A second type of the most frequently employed models consists in *algorithms*. Algorithms represent the phenomenal processes of the different sorts mentioned above as being produced by chains of operations whose succession is necessarily determined -or even in part in a random way- by certain givens and by a system of rules supposed implanted in the agent. These algorithmic models, just like the logico-mathematical systems used in computers, develop purely formal procedures in the sense that they are *neutral as regards content*, at least once a certain level of analysis is reached. They may however take into account the given content belonging to the concrete field in which they are deployed, to the extent that knowledge assimilated by the automaton can be stored. The fundamental problem we are thrown back to here is thus that of the *representation of knowledge*.

It is clear that algorithmic models are reductionist in the weak sense, since they represent processes originally apprehended mentaly by means of abstract logico-mathematical procedures. But this algorithmic reduction which, as mentioned above, is in some way common to all the sciences, must in no way be confused with what I would rather call an "energetic" reduction and which would consist in representing the phenomena in terms of the conflict of forces and equilibrium. The algorithmic models of artificial intelligence clearly belong then to the type that I formerly designated as "cybernetic" or "informational", in which a circulation of information (in the formal sense of the term, it is true) modulates secondary energetic effects such as a displacement or a formulation of an answer. In this sense they are not reductionist.

An ambiguity remains however as to their real aptitude for completely representing the phenomena they intend to represent. Certainly an algorithm is a system of *rules*. But the notion of rule, as understood when applied to mental phenomena, seems to imply the notion of *meta-rule*, from which it seems to me inseparable. And aside from situations of pure calculation, the application of rules is not generally efficient unless controlled by meta-rules whose nature is most often very difficult to formulate.

To what extent can such a notion be adequately transposed into an algorithmic model? It is certainly not the philosopher's role to pronounce interdictions in cases like this; and it is even reasonable to admit that *technical approximations* have been and will continue to be obtained that produce effects quite comparable to those of a true meta-rule. Nevertheless, the philosophical question of the possibility *in principle of a reduction of meta-rules to rules* remains, thus leaving open the question of the success or failure of a reduction, even an algorithmic one, of the features of intelligence. There is one last type of model that may turn out to be particularly fruitful, though it is still quite controversial; it consists in the use of the notion of a "network" originally taken over from anatomy and the physiology of nerve connections. "Connectionism" is based on the double aspect of the combination and randomness network with a multitude of connections, the junctions of which are elementary machines coupled according to different laws and able to vary in accordance with the states of these junctions. The *still confused idea* of "autopoiesis" has been advanced (Atlan, Maturna and Varela, 1975)[4]. Under certain conditions such networks would tend towards a state of equilibrium in the presence of external stimuli. The phenomena of learning and of the recognition of forms would be simulated in this way. It seems that such models as these, being essentially physical in origin, should be reductionist ones. However, the conjecture, that what we are dealing with here is a new and original type of model, neither properly algorithmic nor energetic, and privileging the random factor cannot be disregarded. This would indeed then be a great step not only, as they assure us, towards more and more efficient technical achievements of "intelligent" automata, but also perhaps towards a greater understanding of the specification of "natural" intelligence. It must be recognized, however, that these "connectionist" ideas are still very confused and give rise to extrapolations that are hardly justifiable.

4. CONCLUSION

By way of conclusion, let us briefly sum up the specific problems that seem, from the point of view of reductionism, to concern the models used in artificial intelligence.

This is a discipline that, although still ambiguous in its project and methods, does concern itself in many respects with the study of *human facts*. Taken as a discipline of *scientific knowledge* rather than techniques -even though it is always identified with both- it meets with the same difficulties as does all other social sciences: difficulties inherent in the nature of their object and which seem to rule out any radical reductionism, and perhaps even a moderate one.

It does not seem that the algorithmic formulation of processes, as fruitful as it may be, will ever be quite sufficient. A more vague "heuristic" style of formulation, as in the case of "expert" systems still proves to be indispensable. But this boils down to a merely simplified direct transcription of real human experience, which, though it may *simulate*, *explains* nothing.

The radical difficulty, as I see it, might be expressed as follows: *is it possible to conceive abstract models of the act of symbolization*? The abstract models that figure this act are static. To represent the dynamization of symbolism would be to represent the *creation of symbols* out of a stimulus and to rearrange the old systems and inventing new ones.

When artificial intelligence tackles this problem it seems as though it confuses a genuine solution with the simple consolidation of a stimulus-response schema which is purely "energetic" in nature: a process that no doubt prepares and precedes the symbolic activity but in no way produces symbols. But modelization, which is an eminently intellectual activity, is already in itself a symbolization at a complex level. Can it, as such, adequately achieve its own representation? If the answer to this final question were to be negative, it would signal the existence of a fundamental obstacle to the complete scientific representation of human facts.

If artificial intelligence should be developed as an objective and applied psychology of knowledge and as the adaptation of action to the environment, it can appeal only to types of models that are irreducible to simple physical models, to those we previously called "cybernetic" and "semiotic", and without doubt to still others of the same sort not yet defined. But then it runs up against the radical philosophical obstacle of a *modelization of the symbolic activity*. Let us recognize however that there does not appear to be anything that needs limit its progress *a priori*, as long as we are content with measuring that progress in terms of the perfection of the simulations it manages to attain. This being the case, the philosopher's task should be limited here to exposing and denouncing the ever-present temptation to confuse success in simulation with the reaching of scientific knowledge that would be explanatory of intelligence.

Collège de France

REFERENCE

1.«L'intelligence artificielle comme philosophie», in *Débat*, n° 47, 1987, pp. 88-102.

2. Marcelo Dascal, in *Systems Research*, vol. 6, no 1, pp. 39-52,1982, goes so far as to propose a reformulation of the concept of "knowledge" acapted to the problems of artificial intelligence. («Artificial Intelligence and Philosophy : the Knowledge of Representation»).

3. «Ascribing mental qualities to Machines» in *Philosophical Perspectives of Artificial Intelligence*, Martin Rigle, ed. 1979.

4. Cf. Maturana, H.R. and Varela, F.J., *Autopoiesis and Cognition*, 1980; see also Goschke, Th. and Koppelberg, D., «Connectionism and semantic content of internal representation», in *Revue internationale de Philosophie*, 1/1990, no 171, p. 87

Patrick Suppes

CAN PSYCHOLOGICAL SOFTWARE BE REDUCED TO PHYSIOLOGICAL HARDWARE?

The question of the title I answer in the negative. There are four strands to my argument. The first, which corresponds to the first section of the paper, analyzes the nature of computation. The second concerns the nature of goal-oriented behavior. The third uses an argument that the mind is computationally irreducible. The fourth asserts the irrelevancy of the standard attempts to provide a reduction via general ideas about determinism.

I. Nature of Computation

We may stipulate, I believe, for this paper that the mind is among other things a computational device. This means that matters of computation are of central importance in any arguments about the reduction of psychological concepts to physiological ones. Part of my argument about the irreducibility of computational concepts of the mind to physiological concepts is from the much simpler case of digital computers. In the case of digital computers, we understand to a very much more thorough degree exactly the physical basis of computation and at what level the computational concepts interface the physical concepts. In spite of this great precision of knowledge of interface, we do not at all attempt in the standard theory of computation for digital

E. Agazzi (ed.), The Problem of Reductionism in Science, 183–198.

computers to replace computational concepts by physical ones, which corresponds to replacing psychological concepts by physiological ones. In fact, I am skeptical that even in the case of relatively simple digital computers we could make *direct* physical observations on the computer from which we could infer in detail and without any high degree of error what from a software standpoint was actually being computed. The situation is very much more complicated and difficult, and less likely ever to be understood thoroughly, in the case of our own mental computations, because there is no evidence we will make much headway on the detailed physiological or physical identification of the neurons that are doing any particular computation. Notice of course that this lack of clear identification of physical location, quite apart from understanding the details of that physical location is characteristic of computation in a digital computer. It is a day dream to think that we can easily identify where a particular computation is taking place. Computations in a modern computer move around dynamically. Where they are even placed initially is not a static concept but a dynamic one depending upon what is present and what else is being computed at the time computation is started. It would be a great surprise if something similar is not true of the computations in the brain. Location of mental computations in terms of individual neurons seems totally out of reach. Global location of computations of a particular kind being done in a particular region of the brain is sometimes feasible.

It is also important that physically very different computers compute the same function even by different software programs. Only isomorphism at a high level is usually of interest, and really never in terms of concepts of computation at the level of individual transistors, or to be even more reductionist, at the level of individual elementary particles which make up the many different microscopic parts of a transistor.

Still another concept that seems likely to hold for the brain is that neurons compute statistically, unlike most of the current digital computers. With a statistical computation it is especially unfeasible to think about a reduction from the software to the hardware. Each neuron is making a statistical contribution, but the physical performance of a particular neuron is not of any decisive importance. The feeling is rather here that we have something like the standard result for random variables in probability theory. Given the random variables, which roughly speaking are meant to correspond here to the software concepts, we only have requirements of consistency for there to exist a common sample space. The sample space is never unique. The basic consistency theorems, for example, Kolmogorov's theorem, deal with the existence of an underlying sample space. Without a very large number of assumptions that are not part of the standard theory, there is no unique sample space. The same is surely the situation with the statistical computation of neurons. The underlying neurons actually making the computation can in all likelihood never be observed, –certainly not with present concepts for observation of neurons. A given computation in the brain may be located in different neurons depending upon the dessert one had for lunch or the last green perception that passed through the system. There is such intensive contextualism that a reasonable conjecture is that identifiability for purposes of reduction is out of the question.

In order to eliminate fairy tales, it is important to tighten the argument here and to say that the formal claim of reduction not being possible is relative to a set of observational variables and observational techniques. I will not in this paper attempt such a formalization but I certainly think it is possible and can be done in a straightforward way for simple examples. Obviously, I am not intending to give an *a priori* argument that will hold for all time regardless of what scientific methods are in place a thousand years from now. I am concerned to give at a foundational level an argument in terms of current science and relevant philosophical concepts, an argument that is meant to be a strong one from a computational standpoint, against any possibility of reduction relative to any set of currently observable concepts.

There is one point that might seem elusive in making the distinction between software and hardware in the case of biological organisms as opposed to digital computers. Of course, even in the case of computers, the distinction is not as sharp as it might seem, for in some sense the software program must become a part of the hardware, i.e., a part of the physical organization of the computer. Where, it might be asked, does hardware stop and software begin. Once the software is embodied in the computer, as it is in a different way in the brain, this is not an easy question to answer from purely physical considerations in the case of the computer or physiological ones in the case of the brain. We are able to make the separation in the case of digital computers only in terms of knowledge of what has been done in a deliberate fashion to program the computer, and how the computer hardware has been organized to embody programs. Even this formulation for digital computers is much too simple. First, as already pointed out, the physical location of particular pieces of the program in the memory and processors of the computer is a matter of dynamic allocation and not something we can physically easily directly describe. More importantly, with the current emphasis on digital computers acquiring the capability of learning, the details of the program will not necessarily always be possible to identify.

At least at present our main way of thinking about the brain's software is just in terms of the kind of mental or behavioral concepts psychologists have been developing for a long time and the language of common experience for a much longer time. Moreover, it is undoubtedly these concepts that are the significant targets of any reductionist program. There is no doubt a more detailed and extended sense of software that could be defined. We could include the structural and functional computational changes to be attributed to learning rather than genetic inheritance. Whichever view is taken, reduction to the brain's hardware has no present hope of being carried out in detail.

II. Goal-Oriented Behavior

A second strong argument against reduction of psychology to physiology is to be found in goal-oriented behavior in men and other animals. If I ask my well-trained dog to fetch the newspaper, which has just been delivered, no amount of physiological or physical observation of the dog would be able to either predict the trajectory of his motion as he goes in search of the paper, or infer what task he was attempting to carry out. I am not suggesting that the activity of the mind operates without use of the brain, it is just that psychological concepts cannot be reduced to physiological ones. In other words, we need psychological concepts and the theories that embody these concepts as theories that can be proved independent from a logical standpoint of purely physiological theories. It is a scientific fantasy to think we shall ever be able to make within our present scientific framework sufficient observations on my dog or on any other to determine what task it is engaged in, if those observations are restricted to purely physiological, including neurological, methods of observation and analysis.

To draw a drastic parallel –perhaps too drastic for some-, consider the use of logic in formulating physical theories. No one would be so foolish to say that we can reduce physics or physical concepts to a matter of logic or logical concepts, just because we need logic in the formulation of physical theories. This is how I see the relation of psychology and physiology.To talk about the brain as a machine that we can understand mechanically, which is familiar talk among physiologists and even neurologists, is to talk in a mistaken way. Think how absurd it would be to hear physicists talking about physics as a logical subject and meaning by that that only logic was needed to formulate physical theories. (It is as if logical methods for rigorously proving the independence of axioms or concepts had not been developed for the past hundred years.)

There is another point to be made about goal-oriented behavior. It is easy enough for almost everyone to accept the fact that from purely physical or physiological observations no one can predict where I am going as I leave the house on a complicated physical trajectory to my office, to another house, to a store, or to a restaurant. However, it could be claimed and would be by some die-hard reductionists, that this is simply a case of unpredictable behavior, also to be found in purely physical systems, but unpredictable behavior that can be explained by purely physical concepts. The separation of explanation from prediction is an important matter scientifically but doesn't really bear on the present case, in the following sense. No matter how many observations or how much information of a purely physiological or physical sort is to be collected prior to or after I execute my chosen route to restaurant, store, or whatever, it will be impossible on purely physiological or physical grounds to explain the complicated path I followed. The man on the street recognizes such an enterprise as nonsense. It is important to recognize that from the most fundamental scientific standpoint it is nonsense as well to think in terms of being able to make such a purely physiological or physical analysis of my or any other higher organisms complicated movements.

Physiologists sometimes talk about goal-oriented behavior in cells. Without attempting to judge the scientific merit of this line of analysis, it is evident that we do not have the faintest idea of how to reduce the goal-oriented behavior of a higher organism to goal-oriented behavior of its individual cells. In other words, global goal-oriented behavior cannot be successfully analyzed in terms of goal-oriented behavior of cells. There is a seductive analogue here that could lead to mistaken conclusions. The analogue is that of the reduction of thermodynamics to statistical mechanics. In this case, the behavior of macroscopic parts of matter is reduced at a physical level for certain concepts to the behavior of microscopic particles. Moreover, the relation, as suggested above for neurons in mental computation, is statistical. However, the enormous difficulty of making this reduction a rigorous one, even under the simplest conditions, shows how improbable it is that at any time in the foreseeable future of science we would have the faintest idea of how to carry through a serious reduction of global goal-oriented behavior of an organism to behavior of the organisms individual cells. There is something enormously seductive about this analogue. It is natural to think that we should somehow be able to push through a program of reducing our ordinary behavior as persons to the structure and function of the many billions of cells that make up our bodies. It is a metaphysical dogma that is hard to dislodge. My point is not to say that I can prove it as false, but just to say that the evidence for it is negligible. A way of putting the point is that it is very unlikely that the concepts needed to describe the global behavior of an organism can be reduced to concepts applicable only to individual cells.

It is important to block one mistaken conclusion from the argument I have just given. What I have said is not meant to suggest that we cannot identify at a level even below that of individual cells, for example, at the level of DNA or at the level of genes, microscopic features that predict major features of behavior. The triumphs of genetic analysis of many different sorts of diseases is a major triumph, and something very special about science in this century. It is on the other hand, idle to think that we have even begun to touch the problem of being able to infer goal-oriented movement of an organism from cellular observations. The usual scientific pluralism is at work here. We can do some things well, but not others, in terms of reducing global aspects of behavior to molecular ones. It is a form of metaphysical imperialism —in my view a scientific mistake— to think that we can generalize the successes of molecular biology to carrying out anything like a reduction of all major aspects of goal-oriented behavior. I mention again the difficulties that have been encountered in the last two decades in carrying through in a rigorous way the program of reduction of thermodynamical systems to statistical mechanical ones. It is easy to give from the current literature examples of thermodynamical systems that we do not know how to reduce to statistical mechanical systems. The incomparably more subtle and difficult problem of reduction of the theory of movements of higher organisms seems scientifically totally out of reach.

III. Computational Irreducibility

A familiar important, but not always remarked upon, property of classical physical systems that have been the object of much attention in the history of physics is the property of being computationally reducible. Here is the simplest and most important example. Newton's solution of the two-body problem, i.e., the problem of motion of two bodies acted upon only by the forces of their mutual gravitational attraction, permits us to predict the motion in the future or the past, or the position at any future or past time, given appropriate data on initial conditions at a given time. Moreover, this model has the important property of being applicable in first approximation to two-body systems for the planets, with the sun as one of the bodies. The fact that we can solve the equation of motion in closed form and thereby compute quite directly the future path of the bodies is of fundamental importance. Unfortunately, there was for a long time the feeling that this would be the norm for physical systems, i.e., that most of them were computationally reducible. We would be able to solve the equations of motions to determine the paths of the particles for any indefinite time into the future. However, already in the nineteenth century, the intractable problem of moving from two bodies to three bodies gave plenty of evidence that our ability to computationally reduce most physical systems was probably extraordinarily limited. Moreover, even when we cannot solve the equations of motion in closed form, we often feel that we can do a very good job of a numerical approximation. Already, however, in the case of the three-body problem, as was essentially shown by Poincaré, this was not possible. Now we understand the phenomena very thoroughly for systems that are drastically unstable, as is the situation for some initial conditions in the three-body problem. The numerical methods of computation, necessarily approximate in character, provide a very limited horizon of practical computation concerning the behavior of the system or, to put it in other terms, a very limited horizon of predictability for the behavior of the system. Many other systems of a similar simple physical character have now been identified. With the modern intense interest in chaotic systems we have a sense of limitations in principle about predictability and about computational reducibility of physical systems that did not exist until rather recently, even in so well-established an arena as that of classical mechanics.

There are many reasons to think the mental computations of the mind are also computationally irreducible. One consequence of this is that we shall not be successful in simulating artificially the behavior of the brain. We shall not be successful in the sense that important aspects of human behavior will be missed in any such simulation. Even a model of ten billion artificial neurons will be deficient in providing anything like predictive or computational models. Notice that what we would like is really hopeless: speed up the computations of the brain by four or five orders of magnitude with a model of three or four orders of magnitude less neurons and thereby predict rather well by such computational reducibility future behavior. An unlikely story if ever there was one. To argue that the mind is computationally reducible, as in the case of arguments about physical systems, does not mean that we cannot find subsystems or aspects of behavior that can be computationally reduced. In other words, we can make certain theoretical computations about the behavior of the system that can be verified, and the computation is much simpler and faster than the behavior of the system itself. For example in the case of the three-body problem we can compute for restricted cases the escape velocity of one body, and we can make predictions about the behavior of a qualitative sort without being able to make computations about the detailed behavior. But just as in the case of such physical systems, the brain cannot be computationally simulated in simpler fashion, i.e., be reduced to the computations of the simpler system, when we are concerned with its full behavior. It is a piece of unrealizable scientific fantasy to think that we can move our minds to simulated brains and preserve our psychological identities. There is no reason whatsoever to think such a computational transfer will ever be possible. Above all, the physiologists and neurologists will never make a computational reduction to formulas that lead from individual cell behavior to the mental computations of a fully functioning brain. I am not proposing to offer a metaphysical proof that such a neurological reduction is impossible, it is just that there is no serious scientific evidence whatsoever that it ever will be achievable within the framework of science as we now conceive it.

If I am right in this last claim, it means that in any serious formal or scientific sense reduction of psychological concepts to physiological ones will not be possible. Here what I state informally I mean in a more formal way. Given a formally and empirically adequate psychological theory of psychological phenomena, it will not be possible to prove a representation theorem in terms of a formally and empirically adequate physiological or neurological theory. A certain kind of handwaving may be indulged in by reductionist-minded philosophers, but no serious demonstration of reduction will be given, and it will not be given for substantial reasons. There is no scientific evidence that such a reduction can be carried out at a satisfactory level of detail.

There is still another and different point to be made about computational irreducibility. If we think of having a uniform theory of neurons —meaning that neurons act in the same way from one individual to another and their interaction with software is the same, reduction seems unfeasible. The hopelessness of the situation increases even more when we introduce the hypothesis, which seems likely to have considerable support, that the way in which individual neurons in a given individual interact with software is different from person to person. This would be a natural consequence of biological development occurring in a partially random fashion at the level of dendritic formation and the learning experience in terms of which some of that development is influenced also occurring with random variation from one individual to another. This would mean that the hardware of the neurons is connected to the software of thought in quite different ways in different individuals. If this conjectured variation from individual to individual holds, then reduction is all the more impossible. The detailed structure of computations in one individual, taking the hardware and software together, would differ in quite significant ways from the corresponding structure in any other individual.

IV. Irrelevance of Physical Determinism

The argument for reduction of psychology to physiology, as a byproduct of the reduction of physiology to physics, as a consequence of determinism, is usually not put in as direct and simple a way as I will put it here. I think, however, the force of the argument is as follows. The physical universe is deterministic because as some analytic philosophers, untrained in physics, would put it, it is analytic that like events must have like causes. Given that there is a deterministic account of the physical universe, it then follows that everything that takes place in that physical universe is equally determined. If we know all there is to know about the physical world, that will fix uniquely all the other phenomena including, of course, the mental activities of higher organisms.

Philosophers have been ringing the changes on this argument with different degrees of explicitness and different degrees of emphasis for a long time, at least since the appearance of Kant's *Critique of Pure Reason* in the latter part of the eighteenth century. Kant doesn't discuss explicitly the concept of determinism, for it had not really surfaced in a completely clear way, even though there is a famous passage about the deterministic nature of the universe in LaPlace's introduction to his treatise on probability which appeared not long after the publication of Kant's *Critique*. But there is no doubt that Kant implicitly adopted a deterministic view in his using classical physics in generalized form as the metaphysical foundation of natural science and in his treatment of the category of causality in the *Critique of Pure Reason*.

Of course, Kant was not for a moment prepared to adopt the view that psychology could be reduced simpliciter to physical determinism. He, as in his treatment of the antinomy of free will, was quite prepared to bite the bullet and remove psychology entirely from the domain of science, or as he would put it, more restrictively, natural science. The radical character of Kant's solution to the acceptance of physical determinism, as I would put it, has not always been properly recognized when it comes to working out what one could then hope to do with the science of psychology, but he certainly lays out his views in as explicit a way as could be asked for in the following passage in the preface to the *Metaphysical Foundations of Natural Science*:

> But the empirical doctrine of the soul must always remain yet even further removed than chemistry from the rank of what may be called a natural science proper. This is because mathematics is inapplicable to the phenomena of the internal sense and their laws, unless one might want to take into consideration merely the law of continuity in the flow of this sense's internal changes. But the extension of cognition so attained would bear much the same relation to the extension of cognition which mathematics provides for the doctrine of body, as the doctrine of the properties of the straight line bears to the whole of geometry. The reason for the limitation on this extension of cognition lies in the fact that the pure internal intuition in which the soul's phenomena are to be constructed is time, which has only one dimension. But not even as a systematic art of analysis or as an experimental doctrine can the empirical doctrine of the soul ever approach chemistry, because in it the manifold of internal observation is separated only by mere thought, but cannot be kept separate and be connected again at will; still less does another thinking subject submit to our investigations in such a way as to be conformable to our purposes, and even the observation itself alters and distorts the state of the object observed. It can,

> therefore, never become anything more than a historical (and as such, as much as possible) systematic natural doctrine of the internal sense, i.e., a natural description of the soul, but not a science of the soul, nor even a psychological experimental doctrine. This is the reason why in the title of this work, which, properly speaking, contains the principles of the doctrine of body, we have employed, in accordance with the usual practice, the general name of natural science; for this designation in the strict sense belongs to the doctrine of body alone and hence causes no ambiguity.[1]

Unfortunately, not many philosophers, and I would say almost no scientific psychologists, would be prepared today to follow Kant's way out of the dilemma of determinism.

Of course one immediate response to what I have called the dilemma of determinism is the modern one of saying that quantum mechanics has shown that the microscopic world of physics is not deterministic. I am not going to take that line of argument here because I don't believe it. My own view about quantum mechanics, expressed in several places, is that quantum mechanics is a weak probabilistic theory of the mean.[2] In this view, quantum mechanics is compatible with both deterministic and indeterministic hidden-variable theories, of which perhaps the best example of the latter is stochastic mechanics as developed by Edward Nelson and others, with the understanding that classical Markovian assumptions of Brownian motion must be relaxed to deal with problems of locality. But also as a viable possibility is the kind of deterministic hidden-variable theory outlined by David Bohm and various colleagues. I do not claim fully to understand Bohm's ideas and they are yet to be given an articulated and detailed development, but there is no reason to think that they cannot in principle be elaborated. The difficulty, of course, with any of these extensions of quantum mechanics, in the sense of providing a hidden-variable theory, —one that takes proper account of locality problems—, is being able to make an experimental determination as to whether or not the theory is correct. Such theories may get defeated in the second round by their attempts to go beyond the phenomena of classical quantum mechanics to relativistic particle phenomena, quantum electrodynamics, and more generally to the wide range of experimental findings in elementary particle physics.

What I want to argue is that whoever is right about the proper hidden-variable theory for quantum mechanics, which may turn out to be a purely metaphysical choice, determinism as a general thesis is irrelevant to the question of the reducibility of psychology to physiology. The reason for my holding this view is easy to state. Determinism is too capacious and general a theory to help any such issue to be settled in an interesting way. Why is this? Because the collection of theories that are deterministic is able to accommodate any sort of behavior. Perhaps the way to illustrate this without too many complications and reservations is to consider again the physically simple case of the three-body problem that I have discussed elsewhere with reference to propensity theories of probability.[3] As we move from two-bodies to three-bodies, our detailed understanding of the motions of the three bodies disappears, a fact which I pointed out earlier has been well-known since the nineteenth century. What has not been well-known since the nineteenth century is the proof that for rather simple restricted cases of the three-body problem —meaning a reduction of the problem to the motion of a single-body where the motion of that body is determined by the other two bodies— the following sorts of results hold. First, there exist initial conditions, which in this case are just the initial position in one-dimension and the velocity in one-dimension of the body, such that the sequence of the largest integer values contained in the temporal half-cycles of passing through the plane of the other two bodies has the following property. The sequence of integers so generated, the so-called symbolic dynamics, can represent any random sequence of integers, where the integers are greater than a certain constant. We can therefore represent in terms of the symbolic dynamics of this simple deterministic system -simple in terms of understanding its causes and the derivation of the differential equation governing its motion, not simple in terms of its actual motion-, any random sequence of heads and tails. Second, in contrast there exist initial conditions, in this case just a single number, the velocity in one-dimension of the body, such that the symbolic dynamics encodes the contents of the books in the Library of Congress. One example is purely random, the other is as intentional as you wish, but the richness of this simple, deterministic system is capable of generating either phenomena.

As I have argued in another paper[4] we can develop the same line of attack with purely indeterministic systems. If you don't like determinism choose indeterminism. This choice corresponds to the two choices of hidden-variable theories for quantum mechanics I mentioned above. So does one choose between indeterministic structures of some general probabilistic theory or unstable structures of some deterministic theory? Put in very broad terms the choice seems to be a matter of taste in metaphysics. At the present stage of science there seems no likelihood of any sequence of crucial experiments that will force one of the two positions to the wall. Indeterminism and determinism are here to stay. Exercise your metaphysical choice as you will. There is no inconsistency between determinism and randomness. We can use unstable deterministic systems to generate any probabilistic phenomena desired, or we can take a system that is indeterministic and not known to be deterministic, if that is your metaphysical bent. There are, in fact, some beautiful theorems by Donald Ornstein and his colleagues that make the metaphysical point in still stronger fashion: there are physical systems on which we can make an infinite number of observations —or if you want more precision, there are mathematical models of certain physical phenomena—, such that on the basis of these infinite number of observations it is impossible to distinguish between a deterministic mechanical model governing the phenomena and a stochastic process governing them. This line of argument, the last line of argument I consider here, can be summed up in this way. The classical attempts to reduce psychology or our mental concepts to physical theories and physical concepts by general deterministic arguments is to try to reduce the rich facts of our mental activity to an unverifiable metaphysics of determinism. Kant had the story upside down: our mental life is empirically a rich phenomena which we can study scientifically and successfully. In contrast the general theory of determinism as a view of the universe represents a metaphysics empty of content.

Stanford University (USA)

NOTES

1. *Metaphysical Foundations of Natural Science*, translated by James Ellington, Bobbs-Merrill Company, Incorporated, New York, 1970, pp. 8-9.

2. Probabilistic Causality in Quantum Mechanics. *Journal of Statistical Planning and Inference*, 1990, 25, 293-302.

3. Propensity representations of probability. *Erkenntnis*, 1987, 26, 335-358.

4.Indeterminism or Instability, Does it Matter? In G. G. Brittan, Jr. (Ed.), *Causality, Method, and Modality*, Kluwer Academic Publishers, 1991. Pp. 5-22.

Elena Klevakina

ON THE PROBLEM OF REDUCING VALUE-COMPONENTS IN EPISTEMOLOGY

1. Introduction

In this paper I would like to discuss a problem that I have encountered in working on epistemic concepts (such as knowledge, belief, etc.), but which has some relevance to the theme of reduction. It concerns the value-ladenness of epistemic concepts and, more specifically, the possibility of reducing, i.e. decreasing a value component in them. Before discussing this topic, however, I'd like to clarify what is implied in such expressions as "having an epistemic concept" or "acquiring the concept of belief, knowledge, etc.", in my view. Thus when I say "a concept of knowledge" or a "concept of belief" I do not mean certain mental acts or dispositions to such acts. Moreover, I do not have in mind the particular (or concrete) psychic acts or dispositions but conceptual idealizations of them. That is to say, by knowledge, belief, etc. I mean certain types of universals with a number of intrinsic and necessary features. So in what follows, I will sometimes avoid the term "a concept of belief", for instance, but I will discuss the problem of reducing value-components directly in knowledge, belief, etc..

To see how values come into epistemic concepts, let's consider them, first, from the etymological point of view. Thus, etymologically speaking, epistemic concepts (or, better to say,

E. Agazzi (ed.), The Problem of Reductionism in Science, 199–210.

epistemic terms) spring off from the Greek word "episteme" and, in this sense they are related to knowledge in some respect.

Independant of that, epistemic concepts have to stand in specific relations to knowledge as a goal, for knowledge is, first of all, a goal. (Just consider, for instance, the first sentence of Aristotle's "Metaphysics": "All men by nature desire to know"[1]). In fact epistemic concepts seem to stand in twofold relation to knowledge as a goal. First, they can be means for the goal or they can be directed towards the goal. For example, belief, conjecture or doubt can be means for increasing knowledge or getting nearer to the truth. Second, they can lead away from the truth or knowledge as a goal - this is exactly what error, wrong belief or wrong assertion do.

My argument then runs as follows: standing in certain relations to knowledge as a goal, all epistemic concepts are goaldependent ones. But goal itself is a value. Therefore, all epistemic concepts are value-laden concepts in some sense.

Of course, this thesis has a number of aspects which, however, I wouldn't discuss in this paper. The question which interests me here is as follows: can we reduce or hope to reduce the value-components in epistemic concepts. Thus, in what follows, I want to show that any attempt to reduce the evaluative aspect in them couldn't be completely successful: there are almost always unresolved problems left. At the same time, my point is that the attempted reduction is justified, particularly, for methodological reasons. For, as K. Popper stressed, "we can learn an immense amount even from unsuccessful or incomplete attempts at reduction, and problems left open in this way belong to our most valuable intellectual possessions..."[2]

Having this remark of K. Popper in mind, the central focus of my attention here is concentrated on the question: why we can't succeed as reductionists with regards to the value-components in epistemic concepts.

2. Ways of Reducing the Value-Components in Epistemic Concepts

In my view, there are two possible ways of reduction in question - a "radical" one and a "modest" one.

2.1 What I call a radical way of reducing value components is an attempt to neglect values all together. Being a rather primitive kind of reduction it is still well known from the history of science. Thus while constructing a theory what scholars do sometimes is that they eliminate entities which they do not accept. As an example, take a theory of matter, elaborated by the extreme materialists (L. Büchner, K. Vogt, J. Moleschott), where the category of matter is treated as a primary one, whereas the category of a non-material mind (spirit) is simply eliminated, or in some sense, reduced to the category of matter.

Similarly, the question arises: can we hope to reduce "knowledge", for instance, to a value-neutral concept, and thus, get rid of values in knowledge all together. For the sake of the argument, we can have a brief look at numerous attempts to state necessary and sufficient conditions for someone's knowing a given proposition (P). In fact, in order to give objective criteria of "knowing P", philosophers made different attempts to reduce (in some sense) knowledge to less-evaluated concepts as far as it is possible to do. Accordingly, knowledge was reduced to a "justified true belief", or to a "logically conclusive evidence for P", or to an "adequate evidence for P", etc..[3] Of course, in all these attempts different aspects of "knowing" were under discussion. But what is of interest here is that all philosophical interpretations of knowledge were always rooted in more or less value-laden concepts like "truth", "justified belief", "adequate justification", etc.. Indeed, we prefer a justified belief to a less justified one. But the problem is that what is justified for one person may lack justification for the other. As a result, it may be quite possible for a person to be justified in believing a proposition which is, in fact, false.

The same could be said about "logically conclusive evidence". For in both cases a person may wish or desire to count the evidence as logically conclusive or to count a certain justification as an adequate justification and so on.

Therefore, everything connected with "justification" or "logically conclusive evidence" is a matter of valuation, in my view. Accordingly, having all this in mind, I'm rather in a doubt whether it is possible to achieve a fully satisfactory reduction in epistemic matters if the task is to neglect values all together.

2.2 Let's assume now that it is impossible to get rid of values completely with regards to knowledge, belief, doubt, etc.. Then why not try to decrease a value-component in them. The general way to do this is to trace them back to other concepts with a less evaluative content. As an example of reduction in question, let's take the Aristotelian theory of natural good, that is: "An object (a state of affairs) is a natural good if all persons desire it".[4] Or symbolically,

$$Gx \leftrightarrow (\forall y)\ (y \text{ desires } x)$$

In this case, a value predicate "natural good" is reduced to a less evaluated one "common desire", typical instances of which are "survival", "increasing knowledge according to abilities and interests", etc.. Of course, one cannot say that common desire is completely free from values. Nevertheless, it is less evaluated also from the point of view that it can be investigated empirically. Therefore, in case of the Aristotelian theory the reduction in question results in some kind of a "bridge" (or a "bridge principle") between a value and a fact which, generally speaking, could be confirmed empirically; although one has to admit that the reduction doesn't go through completely in this case.

In my view, analogous reductions could be attempted with regards to epistemic concepts. For instance, why not to consider the possibility of reducing values coming with a certain belief to cognitive values, that is, to those connected with consistency, simplicity, truth, epistemic authority, etc.. Although values are not eliminated here completely, still this could be a step forward towards "purifying" epistemic concepts. Moreover, what is of interest here is how this "modest" kind of reduction could go through.

In fact one can distinguish three types of reduction in question: (a) with regards to a content of belief, i.e. what is believed; (b) with regards to the belief itself, i.e. with regards to the statement as a whole (e.g., "a believes that p"). This isn't already the content of belief, but a mental act of believing; (c) with regards to reasons for believing, i.e. why it is believed.

(a) Let us, first, consider the reduction with regard to the content of belief.

$(a)_1$ One of the examples here could be a scientific belief. Thus when a scientist S believes that "a theory A is better than a theory B" and puts it into more specific words "A is better unified than B" or "A is better confirmed than B", then the content of his belief, i.e. the statement "A is better than B", is already a value-statement, i.e. a statement with a value-predicate in it.

With this fact in mind, one could try to reduce the statement "A is better unified than B" to less value-laden statements such as "A relates phenomena which were not related by B (but also those which B relates)" or "A explains everything from one unifying idea which is not present in B". Or, if we still use the statement "A is the better theory than B" in a different meaning, then we could reduce it to "A is better confirmed than B" which in its turn, could be reduced to "A has passed more severe tests than B", "A is nearer to the truth than B", etc..

Thus, as it seems at first sight, we have reached our aim, that is, we have only cognitive value predicates in the explanatory thesis. However, this is not the case, because there are still hidden value components in the definiens such as "a unifying idea", "relating together", etc.. The crucial point here is that values in question cant't be treated as purely cognitive values. Rather, they are mixed with semantic values (such as clear meaning and maximal truth), methodological values (such as testability and precision), even if these two categories coincide partially with what I call so far cognitive values.

Therefore, with regards to the content of belief shifting from values to cognitive values doesn't help too much. For cognitive values are intertwined with values of another kind.

$(A)_2$ For better confirmation of the thesis stated above, let's make a shift from a scientific belief to beliefs in teleological explanations. The essential point here is that in teleological explanations one explains facts which are not value-laden ones with the help of aims and goals which are certainly value-laden. Accordingly, one has a value-statement in the premise, whereas there's none in the conclusion. Thus,
(1) If x believes that [p is a high value (aim or goal) and q is a necessary condition for obtaining p] , then x behaves in such a way that q obtains.
(2) x believes that [p is a goal and q is a necessary condition for obtaining p].
(3) Therefore, x behaves in such a way that q obtains.

For example, Elizabeth I believed that non-offending the official representatives of the Anglican and Catholic Church (by pronouncing her title unambiguously she thought she would offend them) is a goal and she believed that pronouncing her title ambiguously is a necessary condition to reach the goal. Therefore, Elizabeth I pronounced her title ambiguously.[5] Thus, what we have here is a reduction of a kind which has already been indicated before - a reduction with a "bridge principle" like the Aristotelian definition of "natural good", i.e. of P(x) as a value (goal, aim) to P(x) as a desire, wish, etc.. Observe also, that in teleological arguments the fact explained doesn't (at least apparently) contain values, though they may be hidden. To be more precise, here value statements occur as explanatory statements of a conclusion which does not contain a value-statement: "E uses an ambiguous language". Thus, the reduction goes from the premises to the conclusion. However, in the explanation of the conclusion there are still values.

(b) Reduction with Regards to Belief Itself

My argument so far has been directed to one point - whether it is possible to diminish the value component in epistemic concepts. However, in discussing reduction with regards to belief itself I'd like to reflect on one more point, that is, the consequences of the attempt to reduce.

As a starting point for a discussion I appeal to a principle of epistemic preferability developed by R. Chisholm.[6] In considering the epistemic status of a given proposition h, for S at t Chisholm wants to know what is preferable epistemically: believing h, disbelieving h (i.e. believing the negation of h) or withholding h (i.e. neither believing nor disbelieving h). To summarize the relevant possibilities he uses the following abbreviations: "Bh" for "believing h", "B ¬ h" for "disbelieving h", "Wh" for "withholding h", and "P" for "is epistemically preferable to".

Without attempting to reproduce precisely Chisholm's assumptions which he himself takes as axioms, let us note certain principles which help to exhibit the relations among the concepts stated above:

(a) - If h is true then (Bh)P¬ (Bh), where ¬Bh means not-believing h;
- If h is true then (Bh)P(Wh);
- If h is true then (Bh)P(B¬h);

(b) If (Bh)P (B¬h) then ¬[(B¬h)P(Bh)];

(c) If ¬[(Wh)P(Bh)] then [(Bh)P(B¬h)].

Chisholm deliberately reduces "h is better than y " to " h is preferable to y ". For, according to Chisholm, " h is preferable to y " need not have implications with respect to the goodness or badness of "h" or "y", and "h is epistemically preferable to y " need not have any implication as to the reasonableness or unreasonableness of "h" or "y".[7] With regards to epistemic preferability, Chisholm suggests, it is much more important that the relation P is transitive and the relation P is asymetric.

Interestingly enough, however, Chisholm's principle of epistemic preferability is powerful when limited to belief-relations in epistemic logic, but it is inefficient when applied to the belief itself. Regretably, the consequence of the reduction discussed above is that it makes no room for the psychological process of believing in the sense of investigation into motivation. Accordingly, it leaves behind the so called "why-question", i.e. why withholding h is more reasonable than believing h, and consequently, why believing h is more reasonable than disbelieving h. Our next task, then, will be to analyze the relation of motivation to belief. That is, we now turn from belief itself to reasons for believing.

(c) With Regards to Reasons for Believing

In general, one could discuss the consequences of the reduction in question with respect to two kinds of values, namely, positive cognitive values (like consistency, simplicity, truthlikeness, confirmation, explanatory power, etc.) and negative cognitive values (like non-observability, undefinability, etc.). In this paper I will limit myself to positive cognitive values only. For, in any case, my point is that the degree of valuation in belief can't be reduced to cognitive values exclusively. I will illustrate this point by adhering to consistency and simplicity as cognitive values. The problem is whether these values are by themselves sufficient to determine the belief of an individual scientist in prefering one theory over the other. So, following T. Kuhn, I will consider how consistency and simplicity (as evaluative criteria) functioned in the choice between the heliocentric and geocentric systems.[8] Kuhn stresses that as astronomical theories both Ptolemy's and Copernicus's were internally consistent, but their relation to related theories in other fields was very different. As is known, an essential point of the received theory was the stationary central earth, and the doctrine explained (among other things) how stones fall or how water-pumps function. As to the heliocentric astronomy with its requirements of the earth's motion, it was inconsistent with the existing scientific explanation of those and other terrestrial phenomena. Thus the consistency criterion supported the geocentric tradition.

Simplicity, however, favoured Copernicus, but only if it is evaluated in a special way. On the one hand, if to compare two systems from the point of view of the actual computational labour which is necessary for predicting the position of a planet at a particular time, they proved to be substantially equivalent. Astronomers do a lot of such computations and Copernicus's system didn't offer them no labour-saving techniques, so, it wasn't simpler than Ptolemy's system in this respect. On the other hand, Copernicus'system was simpler if to discuss the amount of mathematical apparatus required to explain not the detailed quantitative motions of the planet, but their essential qualitative features (limited elongation, retrograde motion, and the like) - in this case, Copernicus required only one circle per planet, Ptolemy two. In this sense, the Copernican theory was the simpler and this is the fact which was important to the choices made by both Kepler and Galileo and it was also essential to the victory of Copernicanism. But what is note-worthy here is that sense of simplicity wasn't the only value available nor even the one most natural to professional astronomers whose task was the actual computation of planetary position.

On the basis of the facts mentioned above Kuhn stresses that difficulties in applying standard criteria of choice are typical when scientists must choose between competing theories. And two men having the same criteria for choice may reach different conclusions. For example, they may interpret simplicity differently or have different opinions concerning the range of fields within which the consistency criterion must be met. In any case, even after a detailed comparison of two theories via cognitive values (i.e. consistency and simplicity), in many cases, there remains a lack for the decision, and this lack is filled by a valuation of the scientist, which is not only cognitive any more. Rather, it depends on the goals which a particular scientist has, on his wish to achieve this goal, on his desire to come nearer to the truth, on his specific world outlook, which can be determined by a cluster of cognitive, moral and political values, etc.

3. Conclusion

The foregoing discussion of epistemic concepts has been of largely negative character, being concerned primarily with the problems and difficulties that confront any attempt of reducing value-components in them. However, certain positive conclusions do emerge, in my view.

First of all, we have to meet the fact that knowledge, belief or doubt are intertwined with values which, in the extended sense, consist in or arise from needs, wants and goals we strive for. From this point of view, one could even differentiate the primary type of values and the secondary one. The primary type of value arises in connection with the direct goal, i.e. to achieve knowledge. Secondary values come together with intermediate goals when we are to achieve the primary goal. As a secondary value, one could take rationality, for instance, which could be treated as a method maximizing the efficiency of our attempts to reach the primary goal. In any case, this consideration points out the value-ladenness of belief, knowledge, etc.. as well as the limitations of their "reductive" analysis, i.e. via value-neutral concepts.

At the same time, although it sounds paradoxically, the considerations above also establish the fruitfullness of the method of reduction even with regards to epistemic concepts. For our partial failures reveal some open problems which could appear almost as interesting as their solutions. For example, I do not see how it is possible to "purify" epistemic concepts by keeping out valuations themselves. But why not to have a try to "objectivize" values, i.e. to give them an objective, or let's say, an empirical ground. The problem for methodology, thus, is not whether values are involved into epistemic inquiry, but how they are to be empirically grounded.

An important lesson could also emerge from considering the value relation with regards to this or that epistemic concept. This is the point that there's a good reason for differentiating, so far as it is possible, cognitive, psychological, semantic and methodological values and, accordingly, to estimate the role of this or that value with regards to knowledge, belief, doubt, etc. In any case, if we are to understand human knowledge or human belief, we must be prepared to take account of subtle and delicate relationships between epistemic concepts and values.

Novosibirsk Pedagogical Institute, (URSS)

NOTES

1 Aristotle, *Metaphysics*, p. 1552.
2 K. Popper (1982), p. 133.
3 See for instance Gettier, E. (1963), where he gives a critical account of several attempts to state necessary and sufficient conditions for someone knowing a given proposition.
4 Aristotle, *Nicomachean Ethics*, pp. 1729-1867.
5 See Nagel (1961), pp. 552-553.
6 See Chisholm (1970), pp. 294-301.
7 Op. cit., p. 294.
8 See Kuhn (1977), pp. 323-325.

REFERENCES

Aristotle, "Metaphysics" and "Nicomachean Ethics", in: *The Complete Works of Aristotle*, ed. J. Barnes (New Jersey, 1985).

Popper, K., *The Open Universe* (New Jersey, 1982).

Gettier, E., "Is Justified True Belief Knowledge?", *Analysis* 23 (1963), pp. 121-123.

Nagel, E., *The Structure of Science* (London, 1961).

Chisholm, R., "On a Principle of Epistemic Preferability", *Philosophy and Phenomenological Research*, v. XXX, (September 1969 - June 1970).

Kuhn, T., *The Essential Tension* (Chicago, 1977).

INDEX OF NAMES

Reduction in Science

Structure, Examples, Philosophical Problems

Edited by

Wolfgang Balzer
Universität München

David A. Pearce
Institut für Philosophie, Freie Universität Berlin

and

Heinz-Jürgen Schmidt
Fachbereich Physik der Universität Osnabrück

1984, ISBN 90-277-1811-3
VIII + 446 pp.
Synthese Library 175

Though established as a central issue in contemporary philosophy of science, the theme of reduction has nearly always been approached from an intuitive and informal philosophical perspective, without sufficient attention being paid to fully articulated examples. The main emphasis of the present volume is on the application of formal methods to achieve a deeper understanding of the concept of reduction and to provide a series of precisely reconstructed case studies from different scientific fields, including physics, chemistry, economics and sociology. The wealth of new methods and examples treated here marks a new development in the systematic study of inter-theory relations and the complex process of scientific change.

Audience
The book will be of primary interest to scientists and students in philosophy of science, philosophy, and logic. It will also be of value to scientists and studies in physics, economics, and the social sciences.

Episteme

A SERIES IN THE FOUNDATIONAL, METHODOLOGICAL, PHILOSOPHICAL, PSYCHOLOGICAL, SOCIOLOGICAL, AND POLITICAL ASPECTS OF THE SCIENCES, PURE AND APPLIED

---

1. W.E. Hartnett (ed.): *Foundations of Coding Theory*. 1974 ISBN 90-277-0536-4
2. J.M. Dunn and G. Epstein (eds.): *Modern Uses of Multiple-Valued Logic*. 1977 ISBN 90-277-0747-2
3. W.E. Hartnett (ed.): *Systems: Approaches, Theories, Applications*. 1977 ISBN 90-277-0822-3
4. W. Krajewski: *Correspondence Principle and Growth of Science*. 1977 ISBN 90-277-0770-7
5. J.L. Lopes and M. Paty (eds.): *Quantum Mechanics, a Half Century Later*. 1977 ISBN 90-277-0784-7
6. H. Margenau: *Physics and Philosophy*. Selected Essays. 1978 ISBN 90-277-0901-7
7. R. Torretti: *Philosophy of Geometry from Riemann to Poincaré*. 1978 ISBN Hb 90-277-0920-3 / Pb 90-277-1837-7
8. M. Ruse: *Sociobiology: Sense or Nonsense?* 1979; 2nd ed. 1985 ISBN Hb 90-277-1797-4 / Pb 90-277-1798-2
9. M. Bunge: *Scientific Materialism*. 1981 ISBN 90-277-1304-9
10. S. Restivo: *The Social Relations of Physics, Mysticism, and Mathematics*. Studies in Social Structure, Interests, and Ideas. 1983 ISBN Hb 90-277-1536-X / Pb (1985) 90-277-2084-3
11. J. Agassi: *Technology*. Philosophical and Social Aspects. 1985 ISBN Hb 90-277-2044-4 / Pb 90-277-2045-2
12. R. Tuomela: *Science, Action, and Reality*. 1985 ISBN 90-277-2098-3
13. N. Rescher: *Forbidden Knowledge* and Other Essays on the Philosophy of Cognition. 1987 ISBN 90-277-2410-5
14. N.J. Moutafakis: *The Logics of Preference*. A Study of Prohairetic Logics in Twentieth Century Philosophy. 1987 ISBN 90-277-2591-8
15. N. Laor and J. Agassi: *Diagnosis: Philosophical and Medical Perspectives*. 1990 ISBN 0-7923-0845-X
16. F.P. Ramsey: *On Truth*. Original Manuscript Materials (1927–1929) from the Ramsey Collection at the University of Pittsburgh, edited by N. Rescher and U. Majer. 1991 ISBN 0-7923-0857-3
17. H.A. Shenkin: *Medical Ethics*. Evolution, Rights and the Physician. 1991 ISBN 0-7923-1031-4
18. E. Agazzi (ed.): *The Problem of Reductionism in Science*. 1991 ISBN 0-7923-1406-9

KLUWER ACADEMIC PUBLISHERS – DORDRECHT / BOSTON / LONDON

MIX
Papier aus verantwortungsvollen Quellen
Paper from responsible sources
FSC® C105338

If you have any concerns about our products,
you can contact us on
ProductSafety@springernature.com

In case Publisher is established outside the EU,
the EU authorized representative is:
Springer Nature Customer Service Center GmbH
Europaplatz 3, 69115 Heidelberg, Germany

Printed by Libri Plureos GmbH
in Hamburg, Germany